The Amino Acid SUPER DIET

The Amino Acid SUPER DIET

Gary Zisk, M.D.

G. P. PUTNAM'S SONS
NEW YORK

NOTE TO THE READER

This book should not be regarded, nor is it intended, as medical advice. Before changing your diet, it would be prudent to first consult your physician, particularly one trained and experienced in bariatrics, that branch of medical science dealing with weight reduction.

To Sylvia, Elizabeth and Jeanette

G. P. Putnam's Sons
Publishers Since 1838
200 Madison Avenue
New York, NY 10016

Published simultaneously in Canada

Library of Congress Cataloging-in-Publication Data

Zisk, Gary.
The amino acid super diet.

Includes index.
1. Reducing diets. 2. Amino acids in human nutrition.
I. Title.
RM222.2.Z57 1988 613.2'5 87-19127
ISBN 0-399-13290-2

Book design and composition by The Sarabande Press

Printed in the United States of America

1 2 3 4 5 6 7 8 9 10

Acknowledgments

To The American Society of Bariatrics for the knowledge, training and friendship that helped me to engage in the successful treatment of overweight people.

To Harold Prince, Ph.D., and his staff at DFL Books, for their inestimable help in the preparation of this work.

To Ruth Wreschner, a superb agent, who found me a superb editor, Stacy Creamer.

Contents

PART III

THE COMPLETE AMINO ACID SUPER DIET FOR VEGETARIANS

PART IV

DR. ZISK SOLVES YOUR DIET PROBLEMS

PART I

The First Breakthrough in Reducing Diets in More Than 100 Years

1

Consulting Dr. Zisk

Doctor, why will this diet do more for me than any other diet?

I'd like you to pretend that you're not reading a book, but rather, having a series of consultations with me. Think of yourself sitting comfortably in a chair in my office, finding out everything you should know before you decide to put yourself in my hands. You're a consumer—a purchaser of medical services—and you have the right to ask probing questions. And I, like any doctor, have the responsibility to answer them to the best of my knowledge.

You're here in my office because you want to lose weight as rapidly and painlessly as possible, and you want to keep it off. Maybe you're a veteran of the battle of the bulge. Maybe this is your first serious attempt at getting rid of excess fat. But one thing is certain, you've been bombarded with diet tricks and fad diets and lose-weight-quick schemes.

You probably know that dozens of diet books are published each year. Each introduces a new reducing method or revives an old one. You know that most of the methods work—for awhile. You can take off pounds, but you may not be able to keep them off. The statistics are irrefutable: Nine out of ten people who lose weight on a reducing diet put it back on within a year. Most are fatter one year after dieting than they were before.

So you want a diet that works, not for a few weeks or months, but for a lifetime.

But that's only for starters.

You want more.

You know that, these days, diets often are designed to do more than just get you down to ideal weight. The American Heart Association has a reducing diet that also helps protect against heart attack and stroke. The American Dietetic Association's weight-loss diet helps prevent and manage diabetes. Other diets slim you down and strengthen your immune system, lengthen your life span or make you fit for life.

Up to now, diets have improved your physical condition, and that alone.

The Amino Acid Super Diet improves your *mental* condition as well.

That's what makes it a "super" diet.

As you consult with me, you'll first want to know how this diet helps you lose weight and keep it off. Second, you'll want to know how it helps you become "smarter." Two factors are involved: the reducing factor and the brainpower factor. I'll address these one at a time as I answer dieters' specific questions.

THE REDUCING FACTOR

Doctor, what makes The Amino Acid Super Diet different from all the others?

The Amino Acid Super Diet is based on a series of scientific discoveries made only about a decade ago, which add up to the only *true* advance in weight control in more than 100 years. What they amount to is this: There are certain substances that have the unique property of slimming you down and keeping you slim.

What are these substances?

They're called amino acids.

Are amino acids drugs?

No, they're not. They're ingredients found in everyday supermarket foods. They're nature's own reducing agents.

But aren't acids corrosive? Won't they upset my stomach?

Amino acids aren't strong acids. They're virtually neutral. They are not only safe, they're beneficial.

How do you know a diet based on amino acids works?

Because it has worked in my clinic. Over the last ten years, more than 10,000 patients have gone on my amino acid diet, and more than nine out of ten have successfully lost weight—and kept it off.

Will I like foods containing amino acids?

Of course you will. Some of them are probably among your favorites. There's nothing far out or exotic in this diet. If you like meat, fish, poultry, dairy products, fruits, vegetables, whole grains and baked goods, you like amino acid foods.

I'm a vegetarian. Is this diet for me as well?

There's a version of this diet for vegetarians following the diet and meal plans for nonvegetarians in chapters 6 through 8. See chapters 9 through 11 for details.

How fast can I reduce on your amino acid diet?

As fast as is healthfully possible. Your first, and even your second, week's loss can be around 10 pounds—even more, if you're pretty heavy to start with. But most of that loss is water, not fat. You'll then lose 1 to 3 pounds of fat a week, depending on your weight. The heavier you are, the more you'll lose.

My big failing is that I have no willpower. I can't stay on any diet.

You can stay on this one, because you don't need willpower. The amino acids work automatically. Just follow instructions—eat the foods and take the supplements that provide all the amino acids in the right amounts, and they go to work to turn you from fat to fit. They actually cause flabby fat tissue to be replaced with lean muscle.

But does this diet have all the disagreeable side effects I've experienced on other diets?

Not any of them. That's because amino acids are nature's own reducing agents—they make you feel good, not bad. It's also because you'll be getting your full daily quota of vitamins and minerals, as well as all the other nutrients you need.

What about hunger pangs? I can't stand hunger pangs.

No hunger pangs, and no feelings of deprivation. One wonderful surprise is, you can even have some sweets.

Can you tell me specifically what other disagreeable side effects of dieting are eliminated by your amino acid diet?

No headaches. No nausea. None of that tired, draggy feeling. No circles under your eyes. No hollow cheeks. You won't be embarrassed by bad skin or bad breath. You won't turn people off with your crankiness and irritability. You won't turn from a Jekyll into a Hyde.

I'm a woman, and I have women's problems which affect my dieting. Most diets ignore them. Does yours?

It would be unrealistic to do so. This diet is adjusted to solve your diet problems while you're on the Pill, if you're breast feeding, and during premenstruation, menstruation and menopause. As a matter of fact, the amino acids on this diet can ease the adverse symptoms you may experience during those times of your life.

I can't stay on a diet unless it's simple and it's not way out. Is yours?

The Amino Acid Super Diet is probably the simplest diet ever. You eat the same kinds of foods you've always eaten—only the proportions are different. There's nothing way out involved. No exotic food combinations. No *ugh*-ly health foods. No days on nothing but fruit, or nuts, or rice. No mixing and matching selections from food groups that could take you hours to figure out—*if* you could figure them out. No need to weigh every portion. In fact, you don't have to weigh *any* portion. You don't even have to count calories. Sure, calories count. But I've done all the counting for you. All you have to do is follow an eating plan that you can adjust to your taste as easily as making your choices from a menu.

Let's be practical. Is this diet going to cost me more than I'm now paying for my food? And will I be able to eat out on this diet?

Actually, The Amino Acid Super Diet will probably cost you less. And yes, you certainly can eat out on this diet. I want to make one thing clear: While you're reducing and keeping the pounds off, there will be no disruption of your life-style. But there will be an *improvement* in your life-style—as your brainpower increases.

THE BRAINPOWER FACTOR

Doctor, how does your diet increase my brainpower?

Through the amino acids. There are other nutrients involved in the workings of the brain, particularly vitamins and minerals, but if any nutrients can be called "brain foods," amino acids can. Amino acids make you smart.

Just what do you mean by "smart"?

That's a good question. By smart, I mean having a combination of two qualities: feeling "up" so you can drive to success, and having the mental characteristics to achieve success.

Feeling "up"—isn't that an unusual state to be in? So unusual that people take drugs to get there?

Not at all. It's the natural state to be in. Most of us are so bogged down with cares and worries and responsibilities that we've forgotten what it's like to feel good, to feel wonderful, to feel alive. That's what "up" is. Most of us accept a kind of dull existence—not really depressed, but close to it—as a normal way of life. It shouldn't be. It's not the way nature intended you to feel. Nature intended you to be full of life. Nature intended you to be vibrant and dynamic. You've seen men and women like that—they're rare, but they're around, usually in executive jobs or in starring roles—and you've envied them. You don't have to. Not anymore. The amino acids in your diet give you the "up" nature intended you to have. It's not a drug high, it's the high that comes with sound nutrition.

Just how does feeling up help me drive to success?

When you're naturally up, you're motivated. You don't put things off until tomorrow. You don't sit around and mope. You don't make resolutions you never carry through. How many times have you said to yourself, "Oh, I wish I could change my life. If only I had the mental energy." *Mental* energy. That's what amino acids give you. Think of your brain as a car. The amino acids are the fuel that make it go, that put it in high gear. You can go anyplace you want to go with it. You can go on to success when you have the mental characteristics to drive your brain the right way. And amino acids give you those mental characteristics.

Can you tell me what mental characteristics for success amino acids give me?

Here are just a few. Amino acids help you think more clearly. They cut down mental fatigue (you know what messes you can get into when your brain is overtired). They help you think faster and make the right decisions the first time. They improve your memory and learning skills. They help sharpen your problem-solving skills—and not only for on-the-job problems, but for problems in your home and people problems anywhere. There are many more mental success characteristics that amino acids are responsible for. We'll discuss them as you experience them for yourself as you follow this diet.

Doctor, how do I start on your diet?

Just turn to Week One (page 63 for nonvegetarians; page 123 for vegetarians) and enjoy your first meal on The Amino Acid Super Diet tomorrow morning. It's the only diet that helps you look good, feel good and make good. Then, while the diet is going to work for you, get the answers to more of your questions as you continue to consult with me in the following chapters.

2

Those Amazing Amino Acids: Here's How They Slim You Down

Doctor, how do amino acids help me look good?

By the time you read this, you've probably tried out this diet for a few days, even a week or so, and you've been astounded by the results. You've weighed yourself—naked—each morning before breakfast on your bathroom scale. A consistent daily weighing like this is the most accurate way to keep track of your weight loss. You're probably averaging a drop in weight of at least 1 pound a day, maybe 1½ pounds, maybe more. You're delighted.

Remember, a good deal of that weight loss is excess water you've been carrying around for years because of all that fat.

But water's weight, too—unsightly, flabby weight at that—and you're glad to get rid of it.

You've also begun to feel a wonderful change taking place in how you feel and think. It's as if fat is being trimmed from your brain, and that wonderful organ is becoming lean and athletic—fitter than ever. It's an old cliché to tell you that dieting can make a new you because you'll look so much better. But you've never heard this before: Dieting can make a new you because you'll think so much better.

You're more cheerful, alert, focused and mentally energetic than you can remember being. Life is beginning to feel pretty wonderful.

You've come back to this chapter—to consult with me again—to be sure that these changes that are turning you around, in your body and in your brain, are really going to continue. You know that on other diets, you can lose a lot of water weight fast. You know that fast weight loss makes everybody feel great. And you wonder—do amino acids have anything to do with the new physical and mental you? Because if amino acids don't have anything to do with them, you're going to be in for a rough time as you approach the "danger zone."

The danger zone is the period when the initial water loss stops and your total weight loss comes from burned up fat. Instead of losing 1 pound or more a day, you lose 1 pound or more a *week.* About 3 pounds a week, if you're heavy, is as much as you can shed healthfully. But the sudden change from dropping 7, 10 or more pounds a week to only about 1 to 3 pounds is a shock. And it's discouraging. You know that if the amino acids don't work—if they don't keep your spirits up—your doubts will run amok. You'll say to yourself, "Next week I won't lose anything, and the week after that I'll start *gaining* weight. Oh, what's the use." And you'll give up.

What you want from me is assurance that the amino acids really work, that they will keep your spirits up now that you've entered the danger zone, that you'll be happy to stay with a diet that can slim you down 1 to 3 pounds a week. After all, that's 52 to 156 pounds a year!

I can give you that assurance by answering your questions about *how* amino acids work—how they transform you physically and mentally. Knowing the answers will actually help you diet more efficiently. Many doctors have found, as I have, that when you are convinced that a medical program *can* work, the chances that it *will* work more effectively are dramatically increased. After this consultation, it's possible that you could healthfully lose *more* than 1 to 3 pounds a week.

MEET THE AMINO ACIDS

Doctor, I know amino acids are natural ingredients in foods, but exactly what kind of ingredients are they?

There are twenty-two of them. Twenty are the building blocks of proteins; they are listed in the chart on the following page. Think of the twenty of them as letters in an alphabet. By mixing and matching, you can make virtually an infinite number of words with them. The "words" in this case are proteins. There are hundreds of thousands of different proteins in your body. They build tissues, they make them grow, they repair them, they make almost all of the vital processes of the body work. It could be said that proteins are the stuff of life. And amino acids are the stuff of proteins.

So most amino acids in your diet are effective because they make certain proteins in our bodies. Right?

In our bodies and in our brains, right. But sometimes the amino acids act by themselves. Working either as parts of certain proteins or working by themselves, amino acids slim you down, make you feel up, and stimulate your mental characteristics for success.

THE AMINO ACIDS

Those amino acids that are asterisked can only be obtained from the food you eat. These are called the essential amino acids. The others are manufactured in the body from ingredients in the food you eat, but are also available directly from foods. There are twenty amino acids that make up all proteins.

alanine	*leucine
*arginine	*lysine
asparagine	*methionine
aspartic acid	*phenylalanine
cysteine/cystine	proline
glutamic acid	serine
glutamine	*threonine
glycine	*tryptophan
*histidine	tyrosine
*isoleucine	*valine

Two other amino acids play important roles in the body but are not parts of proteins. They are taurine and ornithine.

Can you tell me which proteins help make your diet work?

There are certain types of proteins in the brain that regulate the way the brain works and, through the brain, how the body works. Some of these proteins control your appetite. Others make you feel not just good, but great. Still others trigger those parts of the brain that help you get the most out of life. Overall, these brain proteins, called "neurotransmitters," can help make you slimmer, happier and smarter. And most of them were discovered within the last ten years.

Can I get these brain proteins directly from protein foods?

No. Because they don't exist in foods of any kind. They have to be made in the brain. What you *can* get from protein foods are the amino acids with which the brain can make these neurotransmitters. So the bottom line is this: To get all the neurotransmitters you need, and get them in the most effective amounts, follow this diet. It supplies the necessary amino acids.

Now I can understand why you say your diet is automatic. The neurotransmitters and the amino acids do all the work for me. Right?

Correct. Now I'd like to tell you just how they do that. Because, as I said before, the more you know, the more belief you'll have in the diet. And the more belief you have in the diet, the better it will work for you. Now I'll answer your questions about the precise way in which amino acids reduce your weight, then how they make you feel up, and finally what characteristics for success they stimulate.

LOOKING GOOD: AMINO ACIDS AS NATURAL WEIGHT-LOSS PRODUCTS

Doctor, do all twenty-two amino acids play a part in reducing my weight?

No. Those that do are phenylalanine, tyrosine, arginine, ornithine, lysine, methionine and tryptophan.

Phenylalanine

Why should I go out of my way to be sure I have the right amount of phenylalanine in my diet?

If you're a compulsive gross overeater, or even if you can't resist eating that little bit more than you should—like gobbling down that extra snack you know you shouldn't have, or sneaking those teensy-weensy pieces off your partner's plate, even though you realize all those teensy-weensy pieces add up to an awful lot of calories—don't rush to a psychotherapist for treatment of your food addiction. Chances are, you're not getting enough phenylalanine in your diet.

If I get enough phenylalanine in my diet, could my compulsive overeating stop?

Yes, it could. The reason is that the right amount of phenylalanine helps manufacture the right amount of the brain protein NE (norepinephrine), the neurotransmitter that helps control appetite. It does this by stimulating the production of still another neurotransmitter, CCK (*c*hole*cy*sto*k*in), the actual brain weight-control protein. When you've consumed just enough nutrients to get down to and maintain ideal weight healthfully, CCK shuts off your appetite as sharply as if you had snapped off a switch.

So with the right amount of phenylalanine in your diet, you don't have to worry about "runaway appetite." You won't have to develop strong willpower to curb your appetite. Phenylalanine—as it is converted in the brain to NE, which in turn stimulates the flow of CCK—curbs it *automatically.*

Every time I have a setback, I get an urge to overeat. Can phenylalanine overcome that urge?

Yes, it can. If you're the kind of overeater who tries to get rid of your daily setbacks, hurt feelings and down periods by cramming your mouth with goodies, phenylalanine can help you—and the millions like you—by getting to the heart of your psychological problem: not being able to cope with the stresses of everyday life without consuming food as if it were a stimulant drug. The NE manufactured by phenylalanine gives you a marvelous feel-good sensation that provides you with the confidence to meet—and beat—everyday stresses. So the psychological compulsion to escape from stress through overeating vanishes.

Doctor, I eat only enough calories to maintain ideal weight for a woman of my age, height and body type. Yet I'm constantly gaining weight. Does lack of phenylalanine have anything to do with it?

It could. Phenylalanine helps manufacture thyroxin, the hormone of the thyroid (it's just below your chin), the gland that regulates just how much of the food you take in is used up. When you don't ingest enough phenylalanine, you don't manufacture enough thyroxin, and your food is not used up but piles up as fat instead. If you "eat like a bird" but are still overweight, lack of sufficient phenylalanine may be the cause.

Tyrosine

I've read that tyrosine is even more effective as a weight-control agent than phenylalanine. Is that true?

Yes, and here's why. On its way to helping create NE in the brain, phenylalanine becomes tyrosine. So tyrosine acts much like phenylalanine, but because it's a step closer to NE, it's a more potent reducing agent.

Why can't I just take tyrosine instead of phenylalanine?

Because you don't "take" tyrosine or phenylalanine or any other amino acid as you take a drug. Amino acids are in the foods you eat; and any one amino acid is not much good unless all the others are present in the right amounts as well. To lose weight, you need all of the seven reducing amino acids in your body/brain, and all the other amino acids as well, and—remember—all in the right amounts. That's what you get on The Amino Acid Super Diet.

Arginine

Why is it most teenagers can eat everything in sight and still stay so slim? I'm no teenager, but I sure would like to know the secret.

It's no secret anymore. The teens are the time of life when the body is brimming with the newly discovered growth hormone. This is a biochemical, secreted by the pituitary gland in the brain, that courses through the body, burning up fat and building up muscle tissue in its place. The growth hormone is created in the brain from the amino acid arginine. Unfortunately, as you grow older, the amount of arginine in your body declines sharply from the teenage high, and so does your natural ability to turn fat into muscle.

Can arginine help turn fat into muscle for me at any age?

Yes, if it's present in the right amount—and you get the right amount on The Amino Acid Super Diet. The growth hormone is stimulated to melt away flabby fat and replace it with toned-up muscle, helping your body to return to teenage fitness.

Teenagers snack a lot. Is it best to get my arginine in the form of snacks?

Only if they're before-bed snacks (you'll learn about the right before-bed snacks on page 162). The reason for this is that the growth hormone is in maximum production during sleep.

Ornithine

I'm glad you told me about how the growth hormone will help me slim down. Is there any way I can be sure my pituitary will produce as much of it it can?

As potent a source of growth hormone as arginine is, it's even more powerful when it works in tandem with the amino acid ornithine. You get the right amount of ornithine in The Amino Acid Super Diet. Remember, too, that all biochemicals manufactured by the body/brain are produced in the right amounts for you *only when your health is good.* The Amino Acid Super Diet contributes to your good health.

Lysine and Methionine

On every diet I've ever been on, I've always suffered—and I mean suffered—from hunger pangs and weakness. I'd like to know how amino acids can do away with that.

The reason the body stores up fats in the first place is to have a supply of food for you in case of famine. That's the way nature intended to keep you from being hungry and feeling weak when you had nothing or almost nothing to eat. So why do dieters, who are eating far more than famine victims, have hunger pangs and feel miserably weak?

The answer is that most reducing diets are so deficient in vitamins and minerals that the stored fats are not burned up properly. Instead of turning into biochemicals the body can use, they turn into biochemicals that create a feeling of weakness. What's more, because these harmful biochemicals can't be used as nutrients, the bottom-line effect is that you're really getting no food from your stored up fat. So you get hungry, very hungry, and it hurts.

Now here's where two amino acids come in to help you. They're lysine and methionine, and they hook up in your body to produce an exciting, newly discovered substance called *carnitine*. What carnitine does is use your stored fats the way nature intended them to be used. That means that when you draw on your body fats to replace the food you're not getting in your meals, your body fats will burn up efficiently, produce the right biochemicals—nutrients your body needs—and you'll feel filled up and filled with energy.

Do lysine and methionine also help reduce weight?

Yes, by producing carnitine. This remarkable substance increases the power of arginine to manufacture the growth hormone responsible for replacing unwanted fat with desirable muscle. Isn't it nice to know that on The Amino Acid Super Diet you get all the lysine and methionine you need to transform yourself from flabby to firm without feeling ravenous or feeble? These two amino acids take a good deal of the suffering out of dieting.

Tryptophan

I'm a sugarholic. I want to ask two questions. First, sugar doesn't contain more calories than the same amounts of other foods—so why does it make me fat? And second, how can I kick the sugar habit?

To answer your first question: Sugar contains no nutritional value at all—no vitamins, no minerals, no amino acids, no fats, no food value, no proteins—nothing. It just supplies calories—"empty calories"—empty of nutrition. Now, your body craves the nutrients that you're not getting when you eat sugar, so for every calorie of sugar you take in, you have to eat at least a calorie of food that contains the nutrients your body needs. Your brain gives you a compulsion to take in those extra calories. And those extra calories mount up, and up, and up—and that's why eating large amounts of sugar makes you fat. Simply stated, as a sugarholic you're compelled by feelings of hunger to eat more—and more and more—than people who are not sugarholics.

How can you kick the sugar habit? With the amino acid tryptophan.

How does tryptophan help me kick the sugar habit?

To answer that, I'd like to start by telling you why you are a sugarholic. It has nothing to do with "a sweet tooth," or a psychological hangup, or a lack of willpower. What it has to do with is a newly discovered substance in your brain, a neurotransmitter called serotonin. When you have enough of it, you feel satisfied after a meal. When you don't have enough of it, you don't feel satisfied. One way to get enough serotonin is to eat sugar—a lot of it. So you're a sugarholic in all likelihood because your brain is deficient in serotonin.

But you can get all the serotonin your brain needs when you get the right amount of tryptophan in your diet. Tryptophan is the essential substance for the manufacture of serotonin in the brain. That's why there's the right amount of tryptophan in The Amino Acid Super Diet to break your sugar addiction.

Doctor, can you quickly summarize how each of the seven weight-reducing amino acids help me slim down?

I've prepared a list for you. Just carry it around and read it several times a day to remind you how amino acids are working to slim you down. Here it is:

SEVEN AMINO ACIDS THAT SLIM YOU DOWN

Phenylalanine helps prevent compulsive eating and psychological addiction to food, and helps regulate your food consumption.

Tyrosine does everything phenylalanine does, but more efficiently.

Arginine burns up fat and replaces it with muscle tissue.

Ornithine improves the weight-reducing property of arginine.

Lysine and **methionine** act together to produce that filled-up feeling and help get rid of unwanted fat.

Tryptophan helps lick the sugar habit, which is a major cause of overweight.

3

More About Those Amazing Amino Acids: Here's How They Keep You "Up" and Help You Succeed

Doctor, how do amino acids help me feel good and make good?

You're probably into the second or third week of the diet by this time, and you're having no trouble at all getting through the "danger zone." Fact is, you're feeling great—but in a very special way.

"You know, doctor," one patient confided, "I've been on

and off diets for heaven knows how long, and come the third week, it's those deprivation blues that get me down. No blues this time. But I'm not just my *regular* self, I feel *better.* A *lot* better."

"In what way better?" I asked.

"Well, I get to work in the morning full of energy, feeling positive about everything, and I stay that way all day long. The people who work with me say, 'What's come over you? Can I get some of it?' But at night after dinner, I feel *sooooo* relaxed, *sooooo* relaxed."

"It's the amino acids at work," I told her. "They can manage your mind and mood. And that means you can manage your life—better."

While I was observing this among my patients over the last ten years, a remarkable woman was proving it in rigorous scientific studies. She is Dr. Judith J. Wurtman, a distinguished research scientist at the prestigious Massachusetts Institute of Technology (M.I.T.) in Cambridge, Massachusetts. Listen to what she has to say:

"Simply put, some foods influence the production and function of chemicals in the brain that are directly involved in determining mood, mental energy, performance and behavior. Some foods increase mental alertness, speed and accuracy and will tend to make you feel more motivated and energetic."

Those are the foods you eat during the day on The Amino Acid Super Diet.

Dr. Wurtman continues: "Other foods are natural tranquillizers that calm feelings of anxiety and stress."

Those are the foods you eat in the evening on The Amino Acid Super Diet.

She adds: "You can . . . make these mood changes occur almost instantly, without investing in long hours of therapy and without drugs—in fact with practically no effort on your part."

All the foods Dr. Wurtman recommends provide the amino acids that are responsible for the right amount of the right neurotransmitters being produced in the brain to control mood. You're enjoying some of those foods on The Amino Acid Super Diet every day; and we'll get to what they are a little later on.

Right now, let's answer your questions about which amino acids influence your moods, and just how they do it. Remember, if you know what's happening to you, why it's happening, and how it can help you, you'll believe in the diet, and your belief will make the diet work better. You'll also be able to let the remarks of skeptics—such as, "Oh, that will never work," or "How many times have you tried and failed? You'll never stick with it"—just roll off like water off a duck's back. Knowledge is strength—the strength to stick with it, because you know you'll come out a winner.

FEELING GOOD: AMINO ACIDS AS NATURAL "UPPERS"

Doctor, what are the substances in the brain that control mood?

They're neurotransmitters, biochemicals that make the brain work. Scientists have discovered fifty-two neurotransmitters in the last ten years or so, but only three of them are associated with mood. When you don't have enough of them, you feel bad, blue, in the dumps. When you have enough of them, you feel up, high, wonderful. You feel nothing's too much for you to handle. You feel great. That's a good way to feel, not only when you're dieting, but anytime.

What are the names of these neurotransmitters that make me feel good?

Two of them are close chemical relatives. They're called NE and DA for short. NE stands for norepinephrine; DA stands for dopamine. The third is serotonin, SE for short.

Can I take these neurotransmitters in the form of pills, and bypass the diet?

If there were such pills—and there aren't—they would be broken up by your digestive system before they could get to the brain. If you injected them into the bloodstream, instead of taking them by mouth, they still couldn't get into the brain. There's a protective wall around the brain—it's called the "blood barrier"—that would keep them out. The only way you can get the feel-good neurotransmitters is from the amino acids in the foods you eat.

What are the amino acids I need to make the feel-good neurotransmitters in my brain?

Phenylalanine, tyrosine and tryptophan.

Aren't those some of the amino acids that slim me down?

Yes. That's why on The Amino Acid Super Diet you feel good while you diet.

If I don't have enough of those amino acids when I diet, how would I feel?

Let's take phenylalanine and tyrosine first. If you don't have enough of them, you won't have enough of NE and DA, and you'll feel a special kind of blues that I call "the jittery blues." You're going to feel grouchy and irritable. You'll feel so tense that at times you'll actually shake. You'll feel restless, jumpy, fidgety. At the same time, you'll feel so depressed that you want to give up on everything. That's no way to feel on a diet.

When I have enough phenylalanine and tyrosine in my diet, how am I supposed to feel?

You've been on the diet for some time now, so let me check off what other patients feel during the day and you tell me if you feel the same way. You feel on top of things, very sharp. You think more quickly, your responses are fast. You feel highly motivated, you want to do things because you feel you can get things done. You're up all day long. Right?

Right! What happens when I don't get enough tryptophan when I diet?

Then you don't get enough SE, and when you don't get enough SE, you get another kind of special blues. I call it "the slow-down blues." You're slow moving, slow thinking. Your reflexes are slow. Your whole body slows down so you're always sleepy, yawny. At the same time, you feel stressed. You feel you have too heavy a load to bear. You wish you could just close your eyes and go to sleep and forget about the whole thing. When you try to get back to yourself, you find you can't concentrate, can't focus in on anything. You feel you're a loser, and you want to give up. That's no way to feel on a diet.

What happens when I do get enough trytophan when I diet?

You've been getting enough on The Amino Acid Super Diet during your evening meal and snacks, so you tell me.

Okay. I feel relaxed. I feel calm. I'm not easily distracted, so when I talk over problems with my family, I can keep my mind on the subject. I'm patient, and that's a surprise because usually I'm not.

And I'll tell you what I'm not feeling. I'm not feeling frustrated anymore. I'm not feeling as if something terrible is going to happen. I feel at peace with myself. Is that the way your other patients feel when they get enough tryptophan?

Yes. Almost all of them. Phenylalanine and tyrosine help build up NE and DA in your brain to give you that active up feeling during the day. Tryptophan is behind the buildup of SE after the evening meal that gives you that calm feeling at night. So these three amino acids help you lick both kinds of depression that plague dieters—the jittery blues and the slow-down blues.

I understand that tryptophan helps me get a good night's sleep. Is that true?

Yes, it is. That's why the snacks on The Amino Acid Super Diet are rich in tryptophan. You won't toss and turn as you did on other diets. You'll sleep soundly and approach the stresses of the next day rested and at ease. That's a good way to travel on the road to success. And amino acids can help you reach the end of the road faster.

MAKING GOOD: AMINO ACIDS AS NATURAL SUCCESS ACTIVATORS

Doctor, I've been following your diet, and I'm now going into the third week. What I'd like to know is,

how can I tell if the amino acids are activating my success characteristics?

You can find out in just about ten seconds by answering the following questions. Please keep in mind that *before* you went on The Amino Acid Super Diet, all the answers probably would have been No. Check that out in your mind as you take the test.

If you checked off just one Yes, that means at least one success characteristic has been improved on The Amino Acid Super Diet. The more Yeses you checked off, the more success characteristics have been stimulated by the diet.

Is there any scientific evidence that success characteristics can be stimulated by amino acids?

Yes, there are hundreds of scientific studies that have established the link between amino acids and success characteristics. Apparently the amino acids work by helping produce neurotransmitters in the brain, which then activate the regions of the brain that determine these characteristics.

Can you tell me just what amino acids are linked to the success characteristics in the quiz I've just taken?

Yes. Just refer to the following table. The numerals before the success characteristics below correspond to those before the questions on the quiz.

I notice that all the amino acids that help slim me down are on the success-characteristic list.

Yes, once again you see the connection between reducing with amino acids and creating a new and better personality for yourself—a happier one and a more successful one. But I

HOW TO TEST WHETHER AMINO ACIDS ARE ACTIVATING YOUR SUCCESS CHARACTERISTICS

	Yes	*No*
1. Do you feel you're more mentally vigorous?	☐	☐
2. Do you feel you're able to handle stress more successfully?	☐	☐
3. Do you feel your decision-making power has been stepped up?	☐	☐
4. Do you feel you can work into the night and still start the next day in peak mental condition?	☐	☐
5. Do you feel that your will to win has been strengthened?	☐	☐
6. Are your feelings about yourself, your work and your goals increasingly positive?	☐	☐
7. Do you feel your memory is improving?	☐	☐
8. Do you feel you can take setbacks with fewer outbursts of irrational anger?	☐	☐
9. Do you feel your learning skills have improved?	☐	☐
10. Do you feel you're now happier at your work?	☐	☐

AMINO ACIDS THAT CAN HELP YOU MAKE GOOD

Success Characteristic	Amino Acids That Stimulate It
1. Mental vigor	cysteine, cystine, glutamine, leucine, lysine, ornithine, phenylalanine, tyrosine, valine
2. Stress control	glycine, methionine, tryptophan, valine
3. Decision-making power	histidine, isoleucine, leucine, threonine, tyrosine
4. Mental-fatigue control	glutamine, methionine, tyrosine
5. Will to win	phenylalanine
6. Positive attitude	glutamine, phenylalanine, tryptophan, valine
7. Heightened memory	glutamine, phenylalanine, taurine, tyrosine
8. Setback management	tryptophan
9. Sharp learning skills	arginine, glutamine
10. On-job happiness	alanine, aspartic acid, glutamine, glycine, serine

want to caution you: There are many factors that shape your personality. Amino acids are just one—up to now a neglected one—and while amino acids certainly can *help* you be the person you want to be, don't forget your genetic heritage and your environment have a lot to do with your personality, too.

The bottom line is likely to be this: Amino acids will help slim you down, and that's a big step toward success. Being fat, for most of us, is not only unhealthy but unsightly in the eye of the beholder, and that's a lot to overcome in the competition for love and jobs. Amino acids will give you the opportunity to make use of positive attitudes instead of negative attitudes—but it's up to you to grasp that opportunity. Amino acids will help produce the neurotransmitters that activate the regions in your brain that stimulate the success characteristics—but it's up to you to take advantage of them.

Amino acids and you: It could be a winning combination.

4

The Best Amino Acid Foods For Dieters—and the Worst

Doctor, what foods should I eat to get the amino acids that control my weight?

There are seven amino acids that can help you reduce. What foods contain them? In devising The Amino Acid Super Diet, that was my problem. I found the answers in the scientific reports of hundreds of researchers. I then knew all of the foods that contain these amino acids, and I selected the best of those foods for my menus—the best, meaning those with the most of these weight-reducing amino acids.

If you're like most dieters, you're anxious to know what those "best" foods are. I asked one patient, "Why?" She answered, "They're magic." Well, they're not. They're good, solid everyday foods you can buy in your supermarket. If there's any magic to them, it's the magic of good nutrition.

Another patient answered, "Because I want to be sure to eat lots and lots of them." Wrong. You mustn't eat too much of any of them. These foods are protein foods, and a high-protein diet may be harmful to some people. What you need is a well-balanced diet with the right amounts of carbohydrates, fats and proteins. The Amino Acid Super Diet is that kind of balanced diet—not a lopsided one. So don't binge on amino acid foods. Follow instructions.

But it is good for you to know what the best amino acid foods for dieters are because, when you recognize them on my menu plan, you'll say, "Ah-ha, these are working for me," and you'll feel confident that you'll lose weight, and lose it naturally, which is the best way to lose weight—the most healthful way.

Doctor, what foods should I eat to get the amino acids that control my weight?

The seven amino acids, you recall, are phenylalanine, tyrosine, arginine, ornithine, lysine, methionine and tryptophan. Here are the best foods for each of them:

You get phenylalanine from low-fat cottage cheese, chicken, beef, eggs, no-fat milk and yogurt, and veal. If you're a vegetarian, you can get this amino acid from almonds, baked beans, peanuts and soybeans.

You get tyrosine from cheese that's been aged, in chicken livers, and in pickled herring. Vegetarians can get it from avocados, ripe bananas, and yeast. Beer and wine are good sources, too, but not—repeat, not—on a diet.

You get arginine and ornithine from chicken, beef, fish, no-fat milk and yogurt, in the animal kingdom. In the plant kingdom there's a good deal of both amino acids in brown rice, whole wheat, sunflower seeds, sesame seeds, oatmeal, nuts and carob. Because carob is a source for two weight-reducing amino acids, it's a wise choice as a chocolate substitute. Instead of putting pounds on you the way chocolate does, carob takes them off.

You get lysine from chicken, eggs, fish and shellfish, and no-fat milk and yogurt. The best way to get it in sufficient quantity from the plant kingdom is by combining garbanzo beans, also called chick peas, with grains, seeds or nuts.

You get methionine in chicken, low-fat cottage cheese, beef, eggs, fish and shellfish, fresh ham, liver, and no-fat milk and yogurt. Vegetarians have to mix legumes with grains, seeds or nuts (but not peanuts, which *are* legumes) in order to get sufficient amounts of this amino acid. Legumes include, in addition to peanuts, dried beans and peas, and lentils; tofu and soy milk are legume products.

You get tryptophan in chicken, beef, eggs, fish and shellfish, and no-fat milk and yogurt. Soybeans are a good source. Elsewhere in the plant kingdom, to get sufficient quantities of tryptophan, you have to combine legumes with grains, seeds, or nuts.

Doctor, I notice that some foods—chicken, for example—contain all seven weight-reducing amino acids. Do they contain other amino acids as well?

That's an astute observation. Yes, they do. These foods are all in the animal kingdom. They contain all the weight-reducing amino acids—five of them are essential amino acids—plus all the other essential amino acids. I like to call them "perfect protein foods." In addition to helping you control your weight, they work to keep you healthy. Without these foods, Dr. Norman Jolliffe, former director of the New York City Department of Health, warned—and I quote—"The reducer may develop weakness, nervousness and fatigue."

PERFECT PROTEIN FOODS FOR NONVEGETARIANS: THEY SUPPLY ALL THE AMINO ACIDS YOU NEED

Asterisked items are richest in amino acids.

*chicken (remove skin before cooking)

chicken livers

cottage cheese, low-fat

*beef, lean

eggs

*fish and *shellfish, especially *sardines

ham, fresh, lean

liver, calves

milk, no-fat

*veal

yogurt, no-fat

All these perfect protein foods are included in The Amino Acid Super Diet eating program for nonvegetarians. They contain all the amino acids that help you reduce and help contribute to your overall good health.

Perfect protein foods are the foundation of The Amino Acid Super Diet. You'll find all of them in the list I've prepared (see page 45).

I'm a vegetarian. Are there perfect protein foods in the plant kingdom?

I'm afraid there aren't. There are a few foods in the plant kingdom that contain all the essential amino acids (but not all the weight-reducing amino acids) but even those foods don't by themselves contain enough of these amino acids to work efficiently in your body and brain.

Then how can a vegetarian go on your diet?

Easily. To get a perfect protein food, all you have to do is *combine* certain plant foods—that is to say, eat them at the same meal. Together, even though they're cooked separately, they supply all the amino acids you need in the right amounts. These "combos" are the basis for the vegetarian version of The Amino Acid Super Diet.

What are the best combos of plant foods to get a perfect protein?

Those that contain either dried beans, lentils and tofu or other soybean products *plus* grain, seeds or nuts. The best combo is dried beans *plus* split peas. I've worked up an interesting variety of combos for you to choose from in the vegetarian version of The Amino Acid Super Diet, and I show you how you can make up combos of your own.

What are the "worst" amino acid foods?

Plant foods—except for certain ones, which I've just mentioned, that are combined to form perfect proteins.

Also, eliminate fatty foods from your diet. They could be changed in the body—"oxidized," as biochemists say—to become poisons that attack amino acids. Another very down-to-earth reason to avoid fatty foods when you're dieting is that they're more calorific than equal amounts of protein or carbohydrate foods. Keep the following foods on your no-no list when you're on The Amino Acid Super Diet. Some nutrition scientists, with prevention of heart attacks in mind, recommend that you keep them on your no-no list at all times.

AVOID THESE FATTY FOODS

cheeses, dessert and hard

beef, prime cuts or fatty choice cuts

cold cuts, except roast turkey (but not turkey roll) and roast chicken, without skin

kidneys and other organ meats, except liver

lamb

milk, whole or 2 percent

pork and pork products, except fresh, lean ham

yogurt, regular

I've heard that some other foods harm amino acids. Is that true?

Yes. Avoid bingeing on a heavy carbohydrate snack. That's to say, a snack that is 80 percent or more carbohydrate in terms of total calories. A huge piece of gloppy chocolate layer cake, for example. Or two pieces. Or three. Or why stop there? Some dieters "fall from grace" from time to time and do just that. But O.D.ing on carbohydrates is no good, because it prevents some amino acids from getting into the brain. The result is that logy feeling—actually a kind of dopiness. Ever feel it?

Any other foods harmful to amino acids?

Here's a surprise. Some sprouts. Sprouts are generally regarded as very healthful, but the sprouts of navy, pinto and kidney beans interfere with the processes in the body that extract amino acids from the proteins in food. So stay away from *those* sprouts.

Let's look at the other side of the picture. Can protein foods be harmful rather than healthful?

Raw egg whites can, for one. While you're not likely to eat them by themselves, you may be fond of some of the foods that have raw egg whites in them. These include—and most people don't know this—meringues, no-bake cakes and pies, chiffon pies, mousses, Bavarian creams, and that very, very high-protein dish, steak tartare. Raw egg whites deplete the body of an important vitamin of the B complex—inositol—and that could lead to your feeling extremely tired all the time, and tossing and turning when you should be sleeping.

The other protein food that can harm you is raw fish. That's also a surprise, because fish is the current "health food of the year." But raw fish can rob your body of two important vitamins of the B complex, B_1 (thiamine) and inositol. That could afflict you with depression, memory loss and inability to concentrate, among other mental and emotional maladies.

What's the best combination of amino acids for me?

All twenty-two—in the right amounts. Some of them you can only get from your food. Some of them are manufactured by your body from ingredients in your food. When your body doesn't manufacture enough of those amino acids, you have to make up the difference with foods containing them. Sounds complicated? Don't worry. On The Amino Acid Super Diet you get all twenty-two amino acids in the right amounts.

Why do you prescribe amino acid supplements?

Because on an 800-calorie diet, there just isn't enough food to supply the right amounts of amino acids. The supplement makes up for the shortages.

One last question. Does cooking protein foods destroy the amino acids?

No. It actually makes the amino acids in the protein foods more available to the body. Cooking also makes protein foods taste better—just compare a raw steak to a charcoal-grilled steak—and more digestable. But keep the heat moderate. High heat "mates" some amino acids, and the couples cling to each other so strongly that digestive enzymes can't split them up so each can go to work.

5

The Right Supplements for Successful Dieting

Doctor, my nutritionist told me I don't need food supplements. Right?

Whether vitamin/mineral supplements are necessary for nondieters has become a major controversy.

On the one hand, traditional nutritionists—these are college-trained professionals, but usually not physicians or Ph.D.s—assert that a "normal" person can get all the necessary vitamins and minerals from foods in a balanced diet selected from the four food groups. Those groups are: animal and plant protein foods; milk and dairy products; grains and products made from them; and fruits and vegetables.

But this doesn't apply to people who are not "normal"—people with ailments, or people recovering from surgery, or people with chronic disorders or who are in chronic pain, or pregnant women or nursing mothers. Think about this for a moment. Put all these people together and they add up to a majority—a substantial majority of the population. These people *need* supplements. And doctors often prescribe them.

There's a body of advanced nutrition scientists—medical experts and Ph.D. researchers—who believe the rest of the population (all the "normal" people) need supplements as well. Their reasoning runs like this: Our food supply has become so processed and contaminated—so *de*natured—that *even* if you select foods from the four food groups, you can't get the right vitamins and minerals in the right amounts. That leads, they hold, to a whole flock of disorders that are not really diseases but rather kind of "diseases in the bud," ranging from mental disorders like depression, anxiety, memory loss, and learning disabilities, to physical disorders, including immune-system impairment, and conditions leading to heart attack, anemia, and a whole catalog of full-blown diseases. One of those diseases, they say, is obesity. Traditional nutritionists don't believe a word of this.

I'm not taking sides until more evidence is in. But one thing is certain: Even with the best of diets, with foods carefully selected from the four food groups, you can't get your full quota of vitamins and minerals and amino acids on a weight-reducing diet of only 600 to 800 calories. Every expert agrees to that, nutritionists and nutrition scientists alike.

The bottom line is: Supplements *are* necessary for dieters.

Doctor, aren't vitamin and mineral pills, and amino acid pills, drugs?

No, they're not—even though they often come in the same forms—tablets, capsules—as drugs. I've already told you that amino acids are beneficial natural substances that the body and brain need. The same is true of vitamins and minerals. Without the right amounts of them, you'll get sick. With severe deficiencies, you'll get very sick. But with the right amounts of vitamins, minerals and amino acids, you could be brimming with life. I think that's an excellent reason to take supplements on a low-calorie diet, don't you?

How do food supplements work on your diet?

As I've said, they make up the shortcomings of a low-calorie reducing diet. Obviously, you need the amino acid supplement to make The Amino Acid Super Diet work. But you need the vitamin/mineral supplement to make the amino acids work. Now I want to stress this: The supplements not only help you to reduce, they help you to feel good, to be healthy—and smarter.

I know that amino acids can affect moods. How about vitamins?

Some vitamins do affect moods—positively. And that's why I supplement The Amino Acid Super Diet so that you get them, and get them in quantities that are effective. Vitamins affect moods by promoting the production of brain neurotransmitters from amino acids. And sometimes vitamins and minerals act on the brain and nervous system directly. Just how each vitamin can help you feel better is spelled out in the following chart.

HOW VITAMINS CAN HELP IMPROVE YOUR MOODS WHILE DIETING

B-complex vitamins:

vitamin B_1 (thiamine) helps keep you free of anxiety and depression

vitamin B_2 (riboflavin) helps fight off stress

vitamin B_3 (niacin) helps you feel positive

vitamin B_6 (pyridoxine) helps you feel "up"

vitamin B_{12} (cobalamin) helps you feel "in control"

choline helps you stay calm

Other vitamins:

vitamin C helps elevate your mood

vitamin E helps keep you mentally sharp

Can minerals also help my moods during dieting?

A few can. Zinc helps you feel mentally alert. Magnesium helps you fight off stress. Calcium helps you feel relaxed.

I've read that mineral supplements also help undo some potential damage due to water loss on diets. Is that true?

Very true. On this diet, you shed an average of 7 pounds or so of water during Week One. That could flush out some water-soluble minerals—potassium, magnesium, selenium, manganese and copper to name a few. You know how important minerals are to your health. So, unless those minerals are replaced, you have a potential health hazard. That's where supplements come to the rescue.

Synthetic versus natural vitamins. What's the difference?

Natural costs more. The difference is significant only in vitamins A, D and B. Vitamins A and D, derived from fish oil, are superior to the synthetic. So are vitamins of the B complex derived from yeast or liver.

Time-release vitamins versus the regular kind. Which is better?

Time-release has the edge. But some vitamin products prescribed to be taken two or three times a day are probably just as effective.

How do I store my supplements?

Keep them in a cool, dark place. Never refrigerate. Here's a helpful hint: Most vitamin containers come with absorbent cotton or a package of moisture absorber. Don't throw either away. By absorbing moisture, they prolong the life of the product. If no moisture absorbers are included, add your own absorbent cotton after you've opened the container.

When do I take my vitamins, minerals and amino acids?

At meals. They're more effective that way. They're also less likely to upset your stomach—if you're one of those people who are susceptible. Take your supplements every day without fail. I said *every* day. When you take supplements, follow the rule of the three C's: continuous, consistent and comprehensive.

Can you explain the rule of the three C's for taking supplements?

Continuous means don't forget to take them and don't deliberately skip a day or two.

Consistent means don't keep experimenting with different combinations. Buy the kind of supplement I recommend and stay with it.

Comprehensive means take *all* the supplements—what I call "complete supplementation." All vitamins and all minerals and all amino acids work together as a team. Repeat: None of these nutrients works as effectively alone. If you take just one single nutrient supplement, or if you take it in less than the recommended amount, you're going to hurt the performance of all the other supplements.

Should I rely on the supplements you recommend for my good health?

That's a very good question, because so many people *do* rely on supplements for their good health. That seems to make them feel they have a license to eat anything they please. I'd like you to take to heart what the late Robert J. Benowicz said. Dr. Benowicz was one of the leading authorities on supplements in the United States. He said:

"Supplementation can support and reinforce good health practices, such as appropriate exercise and an intelligent diet. But relying on supplements alone for good health is absurd. They cannot bestow immunity on a body that continues to be metabolically maltreated and physically abused."

The Amino Acid Super Diet fuses a sound diet with the best selection of supplements. I also recommend exercise. Dr. Benowicz would have liked this approach.

Before I take the supplements you recommend, I want to know if there are any dangers.

There are toxic ceilings for some vitamins and minerals. But reaching them would take an enormous overdose because those ceilings are so far above what I recommend. So, for practical purposes, the supplements are safe.

What supplements do you recommend?

You'll find them listed on the following pages.

A RECOMMENDED AMINO ACID SUPPLEMENT

Dose: Six tablets of an amino acid supplement, containing the dosages shown below, to be taken once a day, with a meal, preferably at breakfast.

Amino Acid	Six tablets supply
alanine	660 mg
arginine	528 mg
aspartic acid	402 mg
cystine	18 mg
glutamic acid	684 mg
glycine	1,650 mg
histidine	48 mg
hydroxylysine	54 mg
hydroxyproline	846 mg
isoleucine	90 mg
leucine	216 mg
lysine	270 mg
methionine	54 mg
phenylalanine	132 mg
proline	984 mg
serine	252 mg

Amino Acid	Six tablets supply
threonine	132 mg
tryptophan	36 mg
tyrosine	18 mg
valine	156 mg

Matching or similar formulations are available at pharmacies and health-food stores.

Note: mg stands for milligrams.

A RECOMMENDED VITAMIN/MINERAL SUPPLEMENT

A (as beta carotene)	25,000 I.U.
(as retinyl palmitate)	5,000 I.U.
B_1 (thiamine HCI)	10 mg
B_2 (riboflavin)	10 mg
niacinamide	100 mg
calcium pantothenate	50 mg
B_6 (pyridoxine HCI)	50 mg
B_{12} (cobalamin concentrate)	30 mcg
folic acid	400 mcg
biotin	100 mcg
C (ascorbic acid)	1,000 mg
D_3 (cholecalciferol)	400 I.U.
E (d-alpha-tocopherol acetate)	400 I.U.
K_1 (phytonadione)	100 mcg
choline (bitartrate)	250 mg
calcium (carbonate)	1,000 mg
magnesium (oxide)	400 mg
zinc (gluconate)	30 mg
iron (ferrous fumarate)	18 mg

Amino Acid	Six tablets supply
manganese (gluconate)	10 mg
copper (gluconate)	3 mg
selenium (Nutrition 21 Yeast)	200 mcg
chromium (Nutrition 21 Yeast)	100 mcg
(as chromium acetate)	100 mcg
iodine (potassium iodide)	150 mcg
molybdenum (sodium molybdate)	150 mcg
magnesium (magnesium trisilicate)	20 mg

Matching or similar formulations are available at pharmacies and health-food stores.

Note: I.U. stands for International Units; mg for milligrams; mcg for micrograms.

PART II

The Complete Amino Acid Super Diet for Nonvegetarians

6

The Nonvegetarian Menu Plan: Week One Getting Off to a Fast Start—Healthfully

Doctor, I just have to lose weight fast. Please help me.

I'd like to tell you about Sally Burns because Sally could be very much like you or like somebody you know. Sally is a modern woman with a bright future ahead of her as a TV producer. But two years ago, her career seemed to be headed for the rocks.

At thirty-four, she had decided to have her first child. Her doctor told her, as most doctors do these days, to put on about 30 pounds during pregnancy. That's a lot, when you consider that back in the 1950s doctors recommended a weight gain of only 10 to 15 pounds. Some women have difficulty shedding today's recommended weight gain, particularly when they don't breast feed, and Sally was one of them.

Actually, she had put on about 25 pounds she couldn't get off after her baby was born. When she got back to work, she was the first to admit she was no longer the svelte, attractive executive who had left her rivals far behind her in the race to the top of the corporate ladder. When she was passed over for a major promotion, she knew she was in trouble.

She wanted to lose weight as fast as possible to make up lost ground, and particularly because the Big Boss was flying in for a meeting with her, and she didn't want him to see her in those clothes for the "full-figured woman." The meeting date was a month away, and could she lose at least 20 pounds by that time? "But I want to do it healthfully," she said. "I must be in top shape for the meeting."

She was. And more than 20 pounds lighter.

Sally Burns isn't alone. The demand for rapid weight loss comes from people in all walks of life for many different reasons. Joe Turner, a career officer in the U.S. Army, faced discharge if he didn't slim down fast. Jane Standish studied her reflection in a full-length mirror and said to herself, "I can never let myself be seen at my daughter's wedding three weeks from now looking like this." Sylvia Patterson, after her husband's death, suddenly found herself reentering the job market and, just as suddenly, faced the need for what she called "instant slimness." Sheila Carpenter said she would "die" rather than let her boyfriend back at college see what she looked like after going home for the holidays.

The hard fact is we're a pretty fat nation—some 60 million of us are overweight. It's no wonder that the demand for quick weight loss is greater than ever. The Amino Acid Super Diet meets that demand. More importantly, it meets it healthfully. Sally said, "I must be in top shape for the meeting." A healthful fast weight-loss diet puts you in top shape for all your meetings—with friends, lovers, relatives—with everybody in your life.

In the preceding paragraphs, I've used made-up names to protect the privacy of people I've treated who were once overweight but aren't anymore. Now I'll be happy to answer the questions you may have about getting off to a fast start on *your* first week on The Amino Acid Super Diet.

Doctor, how much can I lose this week?

On the average around 7 pounds of water and 3 pounds of fat—about 10 pounds.

I'm not "average," doctor, I'm me. Tell me how much I can lose.

All right. But you'll have to do a little arithmetic. How much do you weigh?

185.

Multiply your weight by 15.

That's 2,775.

Correct. Subtract 800.

That's 1,975.

Right again. Now multiply by 7.

13,825?

Right again. Divide by 3,500.

Just about 4?

You've got it! That's the amount of fat *you* will lose on The Amino Acid Super Diet in Week One. Add the 7 pounds of water you'll lose, and your weight loss goes up to 11 pounds. That's losing weight fast.

TO CALCULATE HOW MUCH FAT YOU CAN LOSE IN ONE WEEK:

1. Multiply your present weight by 15.	250 ×15 3,750
2. Subtract the number of calories in your present diet (Week One of The Amino Acid Super Diet).	3,750 −800 2,950
3. Multiply by 7.	2,950 ×7 20,650
4. Divide by 3,500.	5.9 3,500)20,650.0

So the formula seems to say, the heavier I am the more fat I'll lose?

That's correct. If you weigh 250 pounds, the formula works out like this: 250 × 15 − 800 × 7 ÷ 3500 or about 6⅓ pounds. That's a lot of weight, particularly when you'll probably lose more than 7 pounds of water weight in Week One as well. But, remember, as the diet progresses, the lighter you are, the less fat you'll lose.

I read a book by a nutritionist who says it isn't healthful to lose weight fast. The reason she gave is that you have to eat so little food, you don't get enough vitamins, minerals and other nutrients.

She has a point. Because of the way we process our foods these days, the United States Department of Agriculture admits that—and I quote—"In diets of fewer than 1,800 calories, it is hard to get the recommended levels of all essential nutrients." And 1,800 calories a day is a lot of calories. Most women would put on about a pound a week on an 1,800-calorie diet. The U.S. Department of Agriculture goes on to say that shortages of vitamins and minerals in processed foods are particularly acute. So nutrient supplements—vitamins and minerals—are necessary for most of us on even high-calorie diets. As part of this low-calorie diet—about 800 calories a day the first week—I recommend vitamin and mineral supplements be taken to assure that you're getting all the vitamins and minerals you need. That make it healthful. The nutritionist was right when she said that other low-calorie diets that are unsupplemented are unhealthful.

Is this diet also supplemented with amino acids?

Yes. All the essential amino acids in the right amounts to help keep you healthy. And remember, the amino acids also keep you feeling up while you're slimming down. Feeling up, doctors have recently discovered, is a wonderful way to better health. So even though you get a small amount of calories in Week One, you get a large amount of nutrients—nutrients for your better health.

Do you allow junk foods on your diet?

No, I do not. For one thing, they're packed with appetite stimulants—sugar, salts, fats and added chemicals. For another, don't let anybody tell you junk foods are healthful. They're undoubtedly bad for your body, and there's more and more evidence appearing in scientific journals that they're bad for your mental and emotional health as well. Because they're packed with appetite stimulants, junk foods may be one reason there are 60 million overweight people in this country today, and that includes, I'm very sorry to say, an epidemic number of our children. No, junk foods are out—and except, for a treat now and then, they should stay out of your life and your kids' lives forever.

Is Week One a gourmet-diet week?

No. It's a no-frills first week. You want to get the job done. You want to get it done in a way you can handle, eating simple supermarket food, simply prepared. No reason you can't get it at most restaurants. No reason you can't brown-bag it, if you prefer. No reason why you have to spend extra money for it—as a matter of fact, you'll save money. It's good honest food, and you'll like it. But there are rewards coming: In Week Two, *you* will be picking your own favorite foods, and in Week Three, you'll have a taste of sweets and a taste of gourmet dishes.

Any advice before I begin?

Yes, cut down to just the tiniest amount of salt from your salt shaker, and cut out any foods containing salt. (I've prepared a list of such foods for you, which follows.) Reason: cutting way down on salt helps release all that excess water you've been lugging around. Later on, you can get back to a healthful amount of salt. And one final bit of advice: If you're afraid of going on a diet, don't be ashamed of it. A lot of people go on diets for a lot of reasons. Just remember this: Do the thing you fear, and your fear will disappear. Start on The Amino Acid Super Diet first thing in the morning.

AVOID THESE FOODS WITH HIDDEN SALT

- anchovies
- baking mixes, including quick-rising kind
- baking powder
- baking soda
- barbecue sauce
- bouillon cubes
- butter, salted
- buttermilk, salted
- cakes
- candy
- canned foods
- caviar
- celery flakes
- celery salt
- cereals, quick-cooking
- cheeses
- chili sauce
- chinese food
- chocolate, instant mixes
- chocolate milk
- cocoa, instant mixes
- cookies
- corned meats
- delicatessen foods
- finnan haddie
- flour, self-rising
- frozen dinners and prepared dishes
- frozen fish filets
- frozen lima beans
- garlic salt
- gelatin, flavored
- horseradish, bottled
- Japanese foods
- ketchup
- kosher meats
- margarine, salted
- mayonnaise
- meat extracts

meat glazes

meat sauces

meat tenderizers

meats, salted

milk drinks

mineral waters (except low-sodium)

MSG (monosodium glutamate)

mustard

nut butters (except unsalted varieties)

nuts, salted

olives

onion salt

oriental foods, processed

pickles and pickled products

pies

popcorn, salted

potato chips

pretzels

pudding mixes

relishes

rennet tablets

salad dressings

sauerkraut

seasonings with salt added

smoked foods

soft drinks

soups, canned

soups, instant mixes

soy sauce

take-out foods

TV dinners

vegetables, canned

Worcestershire sauce

wursts

THE NONVEGETARIAN DIET (WEEK ONE)

DAY 1

Breakfast

½ medium	grapefruit
1 ounce	dry cereal, with
½ cup	skim milk
	coffee or tea with skim milk

Lunch

½ of 6″	cantaloupe
½ cup	low-fat cottage cheese
1 medium	lettuce, tomato and cucumber salad
1 slice	whole wheat bread
	coffee or tea with skim milk, or diet soda

Dinner

4 ounces	baked flounder
½ cup	cooked carrots
½ cup	cooked squash
½ cup	cooked string beans
	coffee or tea with skim milk

Plus 8 glasses of water per day
Don't forget your daily supplements (see pages 57–59).

THE NONVEGETARIAN DIET (WEEK ONE)

DAY 2

Breakfast

½ cup	tomato juice
1	egg (poached or boiled)
1 slice	toast, with
1 teaspoon	jelly
	coffee or tea with skim milk

Lunch

1 cup	chicken noodle soup (low sodium)
1 3½-ounce can	tuna (water pack, preferably low-sodium, drained), with
1 teaspoon	diet mayonnaise (preferably low-sodium)
1 medium	salad, consisting of lettuce and cucumber, with lemon or vinegar dressing
	coffee or tea with skim milk, or diet soda

Dinner

4 ounces	broiled chicken (skin removed)
½ cup	cooked spinach
½ cup	cooked string beans
½ cup	stewed tomatoes
1 slice	white or rye bread
	coffee or tea with skim milk

Plus 8 glasses of water per day
Don't forget your daily supplements.

THE NONVEGETARIAN DIET (WEEK ONE)

DAY 3

Breakfast

1	navel orange
1 ounce	dry cereal, with
½ cup	skim milk
	coffee or tea with skim milk

Lunch

1 cup	split pea soup (low-sodium)
1	tomato and lettuce sandwich on white toast, with
1 teaspoon	diet mayonnaise (preferably low-sodium)
1 small	pear
	coffee or tea with skim milk, or diet soda

Dinner

8	boiled shrimp, with
1 teaspoon	cocktail sauce
1 medium	tossed salad, with
	lemon or vinegar dressing
½ cup	cooked string beans
½ cup	stewed tomatoes
	coffee or tea with skim milk

Plus 8 glasses of water per day
Don't forget your daily supplements.

THE NONVEGETARIAN DIET (WEEK ONE)

DAY 4

Breakfast

½ cup	tomato juice
1 ounce	dry cereal, with
½ cup	skim milk
	coffee or tea with skim milk

Lunch

2 ounces	turkey white meat, on
2 slices	rye bread
1 medium	tomato and lettuce salad
1 small	apple
	coffee or tea with skim milk, or diet soda

Dinner

4 ounces	broiled flounder
½ cup	cooked string beans
½ cup	cooked carrots
1 medium	lettuce, tomato and cucumber salad
	coffee or tea with skim milk

Plus 8 glasses of water per day
Don't forget your daily supplements.

THE NONVEGETARIAN DIET (WEEK ONE)

DAY 5

Breakfast

½ cup	orange juice
1	egg (poached or boiled)
1 slice	whole wheat toast, with
1 teaspoon	jelly
	coffee or tea with skim milk

Lunch

1	chicken salad sandwich made with
1 teaspoon	diet mayonnaise (preferably low-sodium)
1 medium	salad, consisting of lettuce, tomato and cucumber, with lemon or vinegar dressing
1 small	apple
1 cup	skim milk

Dinner

4 ounces	hamburger (lean ground round), broiled
⅔ cup	cooked string beans
1 small	baked potato (without skin)
	coffee or tea with skim milk

Plus 8 glasses of water per day
Don't forget your daily supplements.

THE NONVEGETARIAN DIET (WEEK ONE)

DAY 6

Breakfast

1	navel orange
½ cup	oatmeal, with
½ cup	skim milk
	coffee or tea with skim milk

Lunch

1 cup	vegetable soup (low sodium)
2 ounces	turkey white meat, on
2 slices	whole wheat or rye bread
1 medium	apple or pear
	coffee or tea with skim milk, or diet soda

Dinner

4 ounces	broiled filet of sole
½ cup	cooked carrots
½ cup	cooked string beans
	coffee or tea with skim milk

Plus 8 glasses of water per day
Don't forget your daily supplements.

THE NONVEGETARIAN DIET (WEEK ONE)

DAY 7

Breakfast

½ cup	tomato juice
1	egg (poached or boiled)
1 slice	toasted protein bread, with
1 teaspoon	jelly
	coffee or tea with skim milk

Lunch

½	apple
½	orange
½	banana
½ cup	low-fat cottage cheese
	coffee or tea with skim milk, or diet soda

Dinner

2 thin slices	broiled calves liver
½ cup	stewed tomatoes
½ cup	cooked string beans
	coffee or tea with skim milk

Plus 8 glasses of water per day
Don't forget your daily supplements.

7

The Nonvegetarian Menu Plan: Week Two Eating More and Losing as Much Fat—With Foods You Choose

Doctor, I don't like some of the foods on the Week One menus. Can I have others?

You are now approaching the "danger zone." This is the week when the big weight loss of Week One tapers off. Your water-weight loss is drying up, and you're only losing fat. This is the time—

May I interrupt, doctor? My nutritionist says I'm also losing some protein, and that's bad.

She's right. So far as other diets are concerned, protein loss is a problem. But *this* diet is supplemented with amino acids, and amino acids are the building blocks of proteins. What happens is this: If any proteins are lost, they're replaced immediately. So, on balance, no proteins are lost. That's another reason this is a healthful diet.

Thank you, doctor. You were saying . . .

I was saying that this time—this "danger-zone" time—is the time most people on *other* diets get discouraged by the slower weight loss and begin to think of giving up. But on *this* diet, the amino acids are working for you. You're not getting dieter's blues. You're not getting physically sick. You're not suffering hunger pangs—you're not suffering at all. As a matter of fact, the amino acids are making you feel high. You feel very positive about the diet. You *want* to continue, and you find you need no willpower to do so.

How does Week Two differ from Week One?

Before I answer that, I'd like you to look over the Substitute Foods lists that begin on page 81. These foods have been selected for their high "nutrient density"—which in plain language means they have about the highest amount of nutrients per calorie of any basic American foods. I'll wait. . . .

Have you looked over the lists?

Yes.

Are there plenty of foods there that you like?

There certainly are.

All right. Now to get back to your question, "How does Week Two differ from Week One?" In Week Two, you have three options:

One: You can repeat Week One as is. No changes. But make your portions a little bigger. That's because recent studies have shown that, since we're only human, if I told you to stay with such extreme restrictions you'd rebel, and eat much more than you should. The fancy name for that is responding to "psychological hunger." Now, with a touch of freedom, and with the amino acids controlling your appetite, you'll probably not eat more than another 200 calories—and that's just fine.

What's the second option? I hope it lets me eat foods on the Substitute Foods lists.

It certainly does. What you can do is substitute any of those foods for any of the foods you don't like on the Week One menus *provided*—and I'm emphasizing this—*provided* both foods, the food you don't like and the food you do like, are on the same numbered list. Say you don't like grapefruit. That's on the Fruit Substitute List—List Number 3. Just run your eye down that list and pick a fruit you *do* like, and substitute it in the amount indicated.

If I decide to go with the Week One menus, substituting for the foods I don't like, can I still eat slightly larger portions, the same as if I was going with the Week One menus as is?

Certainly. I want to give you a sense of freedom. I don't want to prime you to rebel. As I explained, when you have the freedom to eat more, you do. But with the various amino acids curbing your appetite and stimulating your brain, you're not likely to eat more than another 200 calories. And there's certainly nothing wrong with that.

Doctor, you know something—I'd feel really free if I could make up my own menus for Week Two. Any chance?

Yes. That's your third option. You can make up your own 1,000-calorie-a-day menus by following the instructions that follow. (Others like me to do it for them, so there are prepared menus for Week Two beginning on page 92.)

CREATING YOUR OWN WEEK TWO DIET

Make up your meals with the indicated number of servings from the appropriate Substitute Foods lists that follow.

Breakfast	*Number of Servings*
List 3—Fruit Substitutes	1
List 4—Bread Substitutes	1
List 5—Meat Substitutes	1
List 6—Fat Substitutes	1
List 7—Skim Milk Substitutes	1
Lunch	
List 1—Vegetable Substitutes, Raw or	1
List 2—Vegetable Substitutes, Cooked	1

Lunch (cont'd)

List 3—Fruit Substitutes	1
List 4—Bread Substitutes	0 or 1
List 5—Meat Substitutes	1

Dinner

List 1—Vegetable Substitutes, Raw or	1
List 2—Vegetable Substitutes, Cooked	1
List 3—Fruit Substitutes	1
List 4—Bread Substitutes	1
List 5—Meat Substitutes	1

Other permissables—no limits: Coffee and tea, plain; diet soda; unsweetened gelatin. Water, including drinks that are essentially water. You should have at least 8 glasses a day.

Snacks: Any items from List 1, Vegetable Substitutes, Raw—no limits.

SUBSTITUTE FOODS

LIST 1

Vegetable Substitutes, Raw
1 Serving is 1 Cup, Chopped or Sliced
Calories Per Serving: About 20

asparagus

beans, green string, or yellow wax

broccoli

Brussels sprouts

cabbage

cauliflower

celery, celery leaves

cucumber

eggplant

greens: beet, chard, dandelion, chicory, endive, mustard, spinach, turnip

mushrooms

peppers: green, red, sweet, hot

pimento

radishes

salad greens: lettuce, romaine, endive, curly endive, escarole, watercress

scallions, green onions

squash, summer or zucchini

tomato

LIST 2

Vegetable Substitutes, Cooked
1 Serving is ½ Cup
Calories Per Serving: About 40

artichoke, 1 small

beans, green string, or yellow wax

beets

carrots

leeks

onions

peas, green

pumpkin

spinach

squash, summer, winter or zucchini

tomatoes, stewed, or tomato juice, 1 cup

turnip

vegetable or split pea soup, 1 cup (low sodium)

LIST 3

Fruit Substitutes
Servings as Indicated
Calories Per Serving: About 40

apple	1 small
apple juice	⅓ cup
applesauce	½ cup
apricots, fresh	2
cooked	½ cup
dried	4 halves
banana	½
berries:	
strawberries, blackberries	1 cup
blueberries	⅔ cup
loganberries, raspberries	½ cup
cantaloupe	¼ of 6″ melon
cherries	10
cranberry juice	½ cup
dates	2
figs	
fresh	1
dried	1
grapefruit	½ medium
grapefruit juice	½ cup
grape juice	¼ cup

grapes	12
honeydew melon	1/8 of 7″ melon
lemons	2 small
lemon juice	1/2 cup
limes	2
mango	1/4
nectar, canned, apricot, peach, pear	1/3 cup
nectarine	1
orange, navel	1
orange juice	1/2 cup
peach	1
pear	1
pineapple, canned unsweetened	1/2 cup chunks or a small slice
fresh	small wedge
plums	2 small
prune juice	1/4 cup
prunes, fresh or dried	2
raisins	2 level tablespoons (1 ounce)
rhubarb, cooked	1 cup
tangerine	1
watermelon	small wedge

LIST 4

Bread Substitutes
Servings as Indicated
Calories Per Serving: About 70

bagel	½ medium
beans, dried, cooked	½ cup
biscuit	1
bread: white, French, Italian, whole wheat, rye, pumpernickel	1 slice
bread crumbs, dry, grated	¼ cup
soft	⅓ cup
bread sticks	2
bun, frankfurter or hamburger	½
cake, angel food	1 slice, 1½″ at edge
cereal, cooked	½ cup
cereal, dry packaged (non sugared)	
flakes	½ cup
puffed	1 cup
corn, ear	½
corn, whole kernel	⅓ cup

crackers	
graham	2
oyster	½ cup
saltines	5
soda	13
zwieback	2
macaroni, noodles, spaghetti, cooked	½ cup
matzo	5″ square
melba toast	4 rectangles or 8 rounds
muffin, English	½
muffin, plain	1 small
peas, dried, cooked	½ cup
popcorn, popped dry	1 cup
potato, white	1 small
potato, sweet, or yam	¼ cup
rice, cooked	½ cup
roll, dinner	1

LIST 5

Meat Substitutes
Servings as Indicated
Calories Per Serving: About 75

anchovies, no oil	9 thin filets
cheese, American, Cheddar, Swiss	1-ounce slice
clams, oysters, shrimp	4
cottage cheese, creamed	¼ cup
low-fat	3 ounces
crab, lobster	⅓ cup
egg	1
lox, no oil	1 ounce
meat, poultry, liver, fresh fish	1 slice, lean, no bone—cooked weight 1 ounce, raw weight 1⅓ ounces
sardines, no oil	3
scallops	3
tuna (water pack)	½ of 3½ ounce can

LIST 6

Fat Substitutes
Servings as Indicated
Calories Per Serving: About 50

avocado	⅛
bacon, crisp	1 slice
butter or margarine	1 pat (1 teaspoon)
cream cheese	1 tablespoon
mayonnaise, diet	1 level teaspoon
nuts in shell	6
oil for cooking	1 teaspoon
olives	5
salad dressing, French, or oil and vinegar	1 tablespoon

LIST 7

Skim Milk Substitutes
Servings as Indicated
Calories Per Serving: About 80

cottage cheese, dry	½ cup
skim milk, nonfat	1 cup
powdered, dry	⅓ cup
yogurt, plain low-fat	½ cup

Doctor, It's too much trouble for me to sit down and work out a menu plan for myself. Can you work out an 1,000-calorie-a-day menu plan for me and then let me make substitutions from the Substitute Foods lists to eliminate foods I don't like?

Certainly. There are menus for Week Two on the nonvegetarian diet beginning on the following page. You can use them as is, or you can make the substitutions that please you. Be sure you take your vitamin/mineral supplement and your amino acid supplement. You'll note you're getting high amino acid foods in the morning and afternoon for peak energy, mental and physical, and high starch at night to relax you after dinner.

THE NONVEGETARIAN DIET (WEEK TWO)

DAY 1

Breakfast

1 cup	sliced strawberries, or ⅔ cup blueberries
¼ cup	low-fat cottage cheese
½	English muffin, toasted
1 pat (1 teaspoon)	butter
1 cup	skim milk
	plain coffee or tea

Lunch

½ cup	cranberry juice
1 slice	broiled chicken (skin removed)
½ cup	cooked carrots
1 small	salad, consisting of tomato and watercress
	plain coffee or tea

Dinner

½ medium	grapefruit
1 ounce	poached chicken breast
½ cup	cooked macaroni, spaghetti or rice
1 cup	celery
1 cup	cooked cut asparagus
10	cherries
	plain coffee or tea

Plus 8 glasses of water per day
Don't forget your daily supplements (pages 57–59).

THE NONVEGETARIAN DIET (WEEK TWO)

DAY 2

Breakfast

½ cup	orange juice
1	egg (poached or boiled)
1 slice	crisp bacon
1 small	plain muffin
1 cup	skim milk
	plain coffee or tea

Lunch

1 cup	vegetable soup (low sodium)
Up to 5	saltines
½ of 3½-ounce can	tuna (preferably low-sodium, water pack, drained)
1 small	salad, consisting of sliced sweet peppers and mushrooms
½ cup	applesauce
	plain coffee or tea

Dinner

1 ounce	baked flounder
1 cup	shredded cabbage
½ cup	cooked baby onions
2 small	plums
1 slice (1½″ at edge)	angel food cake
	plain coffee or tea

Plus 8 glasses of water per day
Don't forget your daily supplements.

THE NONVEGETARIAN DIET (WEEK TWO)

DAY 3

Breakfast

1	navel orange
¼ cup	creamed cottage cheese
1 slice	whole wheat bread
1 pat (1 teaspoon)	butter
1 cup	skim milk
	plain coffee or tea

Lunch

¼ of 6″	cantaloupe
1 ounce	sliced turkey white meat
1 ounce	Swiss cheese
1 small	steamed artichoke, surrounded by sliced scallions and tomato
2	bread sticks
	plain coffee or tea

Dinner

1 ounce	broiled liver
1 cup	Brussels sprouts, chopped
½ cup	cooked beets
1 small	baked potato
⅔ cup	fresh blueberries
	plain coffee or tea

Plus 8 glasses of water per day
Don't forget your daily supplements.

THE NONVEGETARIAN DIET (WEEK TWO)

DAY 4

Breakfast

½ medium	grapefruit
½ medium	bagel, with
1 tablespoon	cream cheese
1 ounce	lox
1 cup	skim milk
	plain coffee or tea

Lunch

1 cup	split pea or vegetable soup (low sodium)
3	skinless and boneless sardines (preferably low-sodium, packed in water), sprinkled with lemon juice
1 small	salad, consisting of sliced sweet onions and red peppers
1 slice	whole wheat bread
12	grapes
	plain coffee or tea

Dinner

1 cup	popcorn (popped dry)
1 ounce	turkey white meat
1 cup	sliced cucumber
½ cup	cooked carrots or turnip
1 small wedge	watermelon
	plain coffee or tea

Plus 8 glasses of water per day
Don't forget your daily supplements.

THE NONVEGETARIAN DIET (WEEK TWO)

DAY 5

Breakfast

½ cup	grapefruit juice
1	egg (poached or boiled)
½	toasted English muffin or bagel
1 pat (1 teaspoon)	butter
1 cup	skim milk
	plain coffee or tea

Lunch

1 cup	tomato juice
1 small	fruit and vegetable salad of: • romaine lettuce leaves • radishes • ¼ cup creamed cottage cheese • ½ cup canned unsweetened pineapple chunks
	plain coffee or tea

Dinner

⅛ of 7″	honeydew melon
4	boiled shrimp
1 cup	mixture of chopped radishes, cucumber and mushrooms
½ cup	cooked green peas
1 slice	French bread
	plain coffee or tea

Plus 8 glasses of water per day
Don't forget your daily supplements.

THE NONVEGETARIAN DIET (WEEK TWO)

DAY 6

Breakfast

1	navel orange
½ cup	cooked cereal, served with
1 pat (1 teaspoon)	butter
3 ounces	low-fat cottage cheese
1 cup	skim milk
	plain coffee or tea

Lunch

⅛ of 7″	honeydew melon
1 open-faced	chicken salad sandwich, consisting of: • 1 slice whole wheat or rye bread • 1 ounce cubed white chicken meat, combined with • 1 whole canned pimento (preferably low-sodium, packed in water), drained and sliced • ½ cup cooked and chilled green peas • 2 tablespoons chopped celery • 1 teaspoon diet mayonnaise
	plain coffee or tea

Dinner

1 ounce	lean roast beef
¼ cup	mashed yam or sweet potato
½ cup	cooked turnip or winter squash
1 medium	mixed salad, consisting of:
	• 1 small apple, chopped
	• ¼ cup grated carrot
	plain coffee or tea

Plus 8 glasses of water per day
Don't forget your daily supplements.

THE NONVEGETARIAN DIET (WEEK TWO)

DAY 7

Breakfast

1/8 of 7″	honeydew melon
1	egg (poached or boiled)
1 slice	whole wheat bread or pumpernickel
1 tablespoon	cream cheese
1 cup	skim milk
	plain coffee or tea

Lunch

1 ounce	lean hamburger, broiled, on
1/2	hamburger bun
1 small	salad of endive and tomato
1/2 cup	cooked winter squash
1 small	apple
	plain coffee or tea

Dinner

1 ounce	broiled chicken (skin removed)
1/2 cup	cooked spaghetti
1 small	salad, consisting of romaine lettuce, sliced tomato and radishes
1/2 cup	cooked baby onions
1	pear
	plain coffee or tea

Plus 8 glasses of water per day
Don't forget your daily supplements.

8

The Nonvegetarian Menu Plan: Week Three

A Touch of Sugar, a Touch of the Gourmet

Doctor, does a diet have to taste like a diet?

Although you didn't know it during the first two weeks on this diet, you were doing something you never thought you could do—and if I had told you to do it, you would have said you couldn't—but what you've done could change your whole life.

You've broken the junk-food habit.

And you've broken the sugar habit, too.

Actually, the two habits go side by side, because there's

sugar in almost all junk food. Junk food is no good for anybody's health, and it can lead to obesity because it's crammed with appetite stimulants—sugar, salt, MSG, condiments and fats. Sugar is a problem to many overweight people, because it can be habit-forming; and, because it contains zero nutrients, you have to consume more calories of other foods to get the nutrients you need, so you eat twice as much as you should—at least.

As for breaking the junk-food habit, that's great. It's a giant step toward staying slim and healthy for the rest of your life. Instead of junk food, for the last two weeks you've been eating fresh, nutritious, wholesome food, and you've acquired a taste for it. Try this simple test. Pick up some junk food and nibble at it. Surprise—it doesn't taste as good as it used to, does it? A few weeks more on this diet, and it won't taste good at all. Good riddance to junk food!

But, sugar . . . That's another story. Let's not eliminate it entirely. Let me tell you why. It has to do with one of the substances in the brain that help control your appetite—the neurotransmitter serotonin. When you don't have enough serotonin in your brain, you're going to have a craving for carbohydrate foods. Sugar is a carbohydrate food. In one of our previous consultations, I told you how the amino acid tryptophan makes enough serotonin in your brain so that your craving for sugar and other carbohydrate foods is squelched. You feel filled up, and your hunger is turned off. But for some dieters that's not enough. They retain a psychological longing for sugar. I believe that on a diet every longing should be satisfied, otherwise sooner or later you're going to reject the diet. *So, yes, you can have some sugar.*

Sweets on a diet! That's unbelievable! But think of it this way and eating sugar on a diet makes sense: Sugar stimulates the production of serotonin. So if you eat your sweet about 20 minutes before a meal, when you sit down to your meal, the extra serotonoin will have gone to work for you, giving you that filled-up feeling, and your appetite will go way down. Sugar, used in this way, is not an appetite stimulant, it's an appetite suppressant.

Way out? Not at all. The editors of the prestigious *University of California, Berkeley, Wellness Letter* wrote recently, "Including a reasonable amount of sweets on a weight-loss diet may help to ensure that the diet will succeed."

Doctor, what kind of sweets are permitted?

I've prepared a list for you, which appears below. You can have ices, sherbet or ice milk. There are many natural sweets you can have as well.

BEFORE-MEAL SWEETS

apple juice	¼ cup
apricots, dried	2
banana, small	½
carrot, large	1
dates	2
fruit butter	2 teaspoons
ice milk	⅛ cup
ices	⅛ cup
jams or jellies	2 teaspoons
pineapple chunks, canned chunks	⅛ cup
prunes, small	2
raisins	2 teaspoons
sherbets	⅛ cup

I'm glad sweets are back. But, otherwise, does a diet have to taste like a diet?

Of course not. Please understand that in Weeks One and Two my object was to take you off junk food, and the best way to break a habit is cold turkey. That's why you had your fill of nutritious, natural foods prepared simply. But there's no reason why natural foods can't be prepared in more exciting ways. You'll find one such special dish each day on Week Three menus. I call them "touch-of-the-gourmet" dishes.

Just what can I expect on Week Three?

Week Three is essentially Week One with a "touch-of-the-gourmet" dish each day. With these dishes and the before-meals sweets, you'll be eating about 1,000 calories a day.

Do you have to know how to cook to prepare the "touch-of-the-gourmet" dishes?

If you do some cooking, you'll find the recipes quick and easy. Even if you've never cooked, the instructions are so simple, each delicious dish will come out the first time.

I don't cook and I don't want to cook. What are my options?

Ask the cook in the household to prepare the touch-of-the-gourmet dishes for you—or for the whole family. If there's no way at all for you to add these dishes to your menus, then you have two options. One is to repeat Week One, increase portions slightly and add before-meal sweets. The other option is to repeat Week Two, *decrease* portions slightly and add before-meal sweets. But do make an effort to try the "touch-of-the-gourmet" dishes. They're so delicious they'll make the rest of your dieting days so much easier—and they'll become a part of your eating plan when you're down to your desired weight and staying thin.

YOUR OPTIONS FOR WEEK THREE

1. Repeat Week One, but add before-meal sweets and a daily "touch-of-the-gourmet" dish.
2. Repeat Week One, but add before-meal sweets and increase portions slightly.
3. Repeat Week Two, but add before-meal sweets and decrease portions slightly.

Doctor, my choice is 1. But before I go ahead, I'd like to clear up something about the basic idea behind your diet.

Go ahead.

Amino acids come mainly from protein foods, right?

Yes . . .?

Then why are there so many fruits, vegetables, salads and other carbohydrate foods on your diet

There are two reasons.

One is that these carbohydrates are high in fiber. They expand in your stomach, help fill it up and help cut your hunger pangs. Some high-fiber foods also have few or no calories. That's why you can eat all the salad and raw vegetables you like. Just be sure to stay away from commercial dips and salad dressings; they contain many appetite stimulants, and they add unwanted calories. Instead, try a vegetable oil mixed with either vinegar or lemon juice as a salad dressing or a dip.

That makes sense. What's the other reason?

Some of the amino acids that control your hunger cannot get into the brain unless carbohydrates clear the way for them. So, although you would expect an amino acid diet to be a high-protein diet, it isn't. Instead, it's about two-thirds carbohydrate to about one-third protein, because you need about that ratio to get the weight-reducing amino acids into the brain.

Overall, would you say your diet is nutritionally sound?

Let's review. The Amino Acid Super Diet is fairly high in carbohydrates. It has a moderate amount of protein. It's very low in fats—saturated fats (those are the bad ones)—and cholesteral. It's fully supplemented with vitamins, minerals and amino acids. It has no junk foods. I think that adds up to as sound a diet as you can find anywhere.

What do I do once I've completed Week Three of your program but still am not down to my desired weight?

Repeat the Week Three menu plan until you reach the goal weight you've set for yourself. For more about lifetime maintenance, you can consult Chapter 12.

THE NONVEGETARIAN DIET (WEEK THREE)

DAY 1

Breakfast

½ medium	medium grapefruit
1 ounce	dry cereal, with
½ cup	skim milk
	coffee or tea with skim milk

Lunch

½ of 6″	cantaloupe

Replace cottage cheese with Cottage Cheese Supreme (page 113)

1 medium	lettuce, tomato and cucumber salad
1 slice	whole wheat bread
	coffee or tea with skim milk, or diet soda

Dinner

4 ounces	baked flounder
½ cup	cooked carrots
½ cup	cooked squash
½ cup	cooked string beans
	coffee or tea with skim milk

Plus 8 glasses of water per day
Don't forget your daily supplements (pages 57–59).

THE NONVEGETARIAN DIET (WEEK THREE)

DAY 2

Breakfast

½ cup	tomato juice
1	egg (poached or boiled)
1 slice	toast, with
1 teaspoon	jelly
	coffee or tea with skim milk

Lunch

1 cup	chicken noodle soup (low sodium)
1 3½-ounce can	tuna (water pack, preferably low-sodium, drained), with
1 teaspoon	diet mayonnaise (preferably low-sodium)
1 medium	salad of lettuce and cucumber, with lemon or vinegar dressing
	coffee or tea with skim milk, or diet soda

Dinner

Replace broiled chicken with 4 ounces Chicken With Herbs (page 114)

½ cup	cooked spinach
½ cup	cooked string beans
½ cup	stewed tomatoes
1 slice	white or rye bread
	coffee or tea with skim milk

Plus 8 glasses of water per day
Don't forget your daily supplements.

THE NONVEGETARIAN DIET (WEEK THREE)

DAY 3

Breakfast

1	navel orange
1 ounce	dry cereal, with
½ cup	skim milk
	coffee or tea with skim milk

Lunch

Replace split pea soup with 1 cup Puree Tomato-Pea Soup (page 115)

1	tomato and lettuce sandwich on white toast, with
1 teaspoon	diet mayonnaise (preferably low-sodium)
1 small	pear
	coffee or tea with skim milk, or diet soda

Dinner

8	boiled shrimp, with
1 teaspoon	cocktail sauce
1 medium	tossed salad, with lemon or vinegar dressing
½ cup	cooked string beans
½ cup	stewed tomatoes
	coffee or tea with skim milk

Plus 8 glasses of water per day
Don't forget your daily supplements.

THE NONVEGETARIAN DIET (WEEK THREE)

DAY 4

Breakfast

½ cup	tomato juice
1 ounce	dry cereal, with
½ cup	skim milk
	coffee or tea with skim milk

Lunch

2 ounces	turkey white meat, on
2 slices	rye bread
1 medium	tomato and lettuce salad
1 small	apple
	coffee or tea with skim milk, or diet soda

Dinner

Replace broiled flounder with Tangy Broiled Flounder (page 116)

½ cup	cooked string beans
½ cup	cooked carrots
1 medium	lettuce, tomato and cucumber salad
	coffee or tea with skim milk

Plus 8 glasses of water per day
Don't forget your daily supplements.

THE NONVEGETARIAN DIET (WEEK THREE)

DAY 5

Breakfast

½ cup	orange juice
Replace egg (poached or boiled) with Chinese Omelette (page 117)	
1 slice	whole wheat toast, with
1 teaspoon	jelly
	coffee or tea with skim milk

Lunch

1	chicken salad sandwich made with
1 teaspoon	diet mayonnaise (preferably low-sodium)
1 medium	salad, consisting of lettuce, tomato and cucumber, with
	lemon or vinegar dressing
1 small	apple
1 cup	skim milk

Dinner

4 ounces	hamburger (lean ground round), broiled
⅔ cup	cooked string beans
1 small	baked potato (without skin)
	coffee or tea with skim milk

Plus 8 glasses of water per day
Don't forget your daily supplements.

THE NONVEGETARIAN DIET (WEEK THREE)

DAY 6

Breakfast

1	navel orange
Replace oatmeal with	Oatmeal Delight (page 118), with
½ cup skim milk	
	coffee or tea with skim milk

Lunch

1 cup	vegetable soup (low sodium)
2 ounces	turkey white meat, on
2 slices	whole wheat or rye bread
1 medium	apple or pear
	coffee or tea with skim milk, or diet soda

Dinner

4 ounces	broiled filet of sole
½ cup	cooked carrots
½ cup	cooked string beans
	coffee or tea with skim milk

Plus 8 glasses of water per day
Don't forget your daily supplements.

THE NONVEGETARIAN DIET (WEEK THREE)

DAY 7

Breakfast

½ cup	tomato juice
1	egg (poached or boiled)
1 slice	toasted protein bread, with
1 teaspoon	jelly
	coffee or tea with skim milk

Lunch

½	apple
½	orange
½	banana
½ cup	low-fat cottage cheese
	coffee or tea with skim milk, or diet soda

Dinner

2 thin slices	broiled calves liver
½ cup	stewed tomatoes
½ cup	cooked string beans

Replace coffee with skim milk with Café au Lait (page 119)

Plus 8 glasses of water per day
Don't forget your daily supplements.

"TOUCH-OF-THE-GOURMET" DISHES

Cottage Cheese Supreme

½ cup canned unsweetened or fresh pineapple chunks
3 ounces low-fat cottage cheese
½ cup low-fat plain yogurt
Dash ground cloves
⅛ teaspoon ground cinnamon
1 package Equal (optional)
3 tablespoons unsweetened pineapple juice
½ teaspoon fresh lemon juice
1 teaspoon regular wheat germ
1 tablespoon raisins
1 tablespoon coarsely chopped walnuts

1. Drain pineapple chunks, reserving juice. Place in cereal bowl. Spoon cottage cheese over pineapple.
2. Place yogurt in measuring cup. Stir in spices. Dissolve Equal in pineapple juice and add lemon juice. Add to yogurt, and gently stir to blend. Pour over cottage cheese.
3. Top with wheat germ, raisins and nuts. Serve well chilled.

Yield: Serves 1

Chicken With Herbs

1 tablespoon minced garlic
2 tablespoons minced shallot or onion
2 tablespoons Italian olive oil
3 tablespoons apple cider vinegar
1 tablespoon balsamic vinegar
¼ teaspoon salt
½ teaspoon each dried crushed rosemary and chili con carne seasoning
1 teaspoon dry mustard
Dash cayenne
1 3-pound broiling chicken, cut into eight pieces, skinned
Minced fresh parsley

1. Prepare marinade by combining in a jar all ingredients except the chicken and parsley. Cover, and shake well to blend.
2. Wash chicken and pat dry with paper toweling. Prick all over with sharp-pronged fork. Place in deep bowl. Spoon marinade over chicken, turning pieces to coat them. Cover and refrigerate for 3 hours or longer. Remove from refrigerator 1 hour before cooking.
3. Preheat broiler. Drain chicken, reserving marinade. Place in lightly oiled shallow broiling pan. Broil 6″ from heat for 8 minutes. Turn, spoon with marinade, and broil for 8 minutes. Continue turning, spooning with marinade, and broiling until chicken is done (about 30–35 minutes). Sprinkle with parsley and serve.

Yield: Serves 4

Pureed Tomato-Pea Soup

2 tablespoons Italian olive oil
1½ cups coarsely chopped onion
1 tablespoon each coarsely chopped garlic and celery
1 tablespoon apple cider vinegar
2½ pounds fresh ripe tomatoes, skinned, cored and chopped
¾ cup split peas, washed
2 cups chicken broth
½ cup tomato puree
2 tablespoons coarsely chopped fresh basil
¼ teaspoon dried thyme leaves, crushed
¼ teaspoon salt
1 bay leaf
1 tablespoon curry powder
Minced fresh parsley

1. Heat oil in stainless-steel pot or waterless cooker. Add onion, garlic and celery, and sauté until wilted. Stir in vinegar, and cook for 1 minute.
2. Add balance of ingredients, except parsley. Bring to boil. Reduce heat. Cover, and simmer for 1¼ hours, stirring from time to time. Remove from heat. Partially uncover, and let cool briefly in pot. Discard bay leaf.
3. Pour into food blender or processor, and puree. Reheat, if necessary, and serve sprinkled with parsley.

Yield: Serves 6

Tangy Broiled Flounder

½ cup fresh mint leaves, rinsed, dried and minced
3 tablespoons apple cider vinegar
1 tablespoon balsamic vinegar
¼ cup unsweetened apple juice
4 whole baby flounder (¾ pound each), cleaned
4 teaspoons peanut or corn oil
¼ teaspoon salt
2 tablespoons each minced shallot and sweet red pepper
2 teaspoons minced garlic

1. Combine first four ingredients in a jar. Cover, and shake well to blend. Let stand at room temperature for 2 hours or longer.
2. Preheat broiler. Wash fish and pat dry with paper toweling. Brush with oil; sprinkle with salt. Scatter half of shallot, sweet red pepper and garlic mixture over fish. Broil 6″ from heat for 8 minutes.
3. Spoon 1 tablespoon of mint mixture over each fish. Return to broiler for 1 minute. Turn carefully. Sprinkle with remaining shallot, sweet red pepper and garlic. Broil for 8 minutes. Spoon 1 tablespoon mint mixture over each fish. Broil until done. Fish should flake easily when tested with fork; do not overcook.
4. Serve on warmed individual plates with pan juices in sauceboat.

Yield: Serves 4

Chinese Omelette

For the filling:

2 teaspoons peanut oil or unsalted Oriental sesame oil
1 small onion, thinly sliced
1 whole scallion, diagonally cut into ½″ pieces
¼ teaspoon each peeled and minced fresh ginger and garlic
1 small tofu cake (bean curd), drained on paper toweling
2 dashes cayenne
½ teaspoon reduced-sodium soy sauce

For the eggs:

4 eggs
⅛ teaspoon salt (if desired)
2 tablespoons evaporated milk
¼ teaspoon peanut oil or unsalted Oriental sesame oil

1. To prepare filling, heat 2 teaspoons oil in 8″ nonstick skillet until hot but not smoking. Add onion, scallion, ginger and garlic to pan, and sauté for 2 minutes over medium heat. Add tofu. Sprinkle with cayenne and soy sauce. Stir and cook until well heated. Transfer to a small bowl and cover to keep warm. Wipe out skillet.
2. In a small bowl, combine eggs with salt and milk, beating lightly with fork to blend.
3. Brush skillet with ¼ teaspoon oil, and reheat. Add eggs, tilting pan so that mixture spreads evenly across skillet. Cook until lightly brown (cook on one side only). Slide onto warm serving plate. Spoon vegetable mixture over half of omelette. Flip over other side. Cut omelette in half, and serve.

Yield: Serves 2

Oatmeal Delight

1 tablespoon apple or orange juice concentrate
⅔ cup cold water
⅛ teaspoon each salt and ground ginger
2 dashes ground allspice
⅓ cup old-fashioned rolled oats
Ground cinnamon to taste

For a smooth-textured cereal:

1. Combine all ingredients except cinnamon in small saucepan. Bring to boil. Turn heat down, and gently simmer for 8–10 minutes, stirring often.
2. Cover and let stand for 2 minutes.
3. Pour into serving bowl, sprinkle with cinnamon, and serve.

For a thicker, more chunky cereal:

1. Combine juice, water, salt, ginger and allspice in small saucepan. Bring to boil. Stir in oatmeal. Bring to boil again; reduce heat, and gently simmer for 8–10 minutes, stirring often.
2. Cover and let stand for 2 minutes.
3. Pour into serving bowl, sprinkle with cinnamon, and serve.

Yield: Serves 1

Café au Lait

1/3 cup equal parts water and apple juice, no sugar added
1/2 teaspoon decaffeinated coffee
1 teaspoon grain coffee
1/8 teaspoon ground cinnamon
1/2 cup skim milk

1. Combine first four ingredients in small saucepan and bring to boil, stirring to blend.
2. Reduce heat to simmering. Add milk and reheat until hot (do not boil).

Yield: Serves 1

PART III

The Complete Amino Acid Super Diet for Vegetarians

9

The Vegetarian Menu Plan: Week One More Food, But No Added Calories

Doctor, can I enjoy all the advantages of The Amino Acid Super Diet on the Vegetarian version?

There have been many reducing diets that are essentially vegetarian—and there are good reasons for that. Calorie for calorie, you get more bulk from plant food than you get from animal food, and that helps your stomach feel filled up. The fiber—and there's a lot of it in a vegetarian diet—helps, too. It swells up in the stomach.

There's also very little fat in plant food, and ounce for ounce, fat is more than twice as caloric as other basic nutrients. So you can eat more plant food than animal food without adding a single calorie. It's no wonder that vegetarian diets succeed.

But this vegetarian diet is even better than the others because you can enjoy all the advantages of the nonvegetarian version of The Amino Acid Super Diet as well. You'll get the same mental and emotional lift while you slim down, and your first-week weight loss will be just as rapid—about 10 pounds of fat and water.

If you turned to this chapter first to get started, it would be a good idea to read what I have to say about the nonvegetarian diet in the preceding chapters. All the good things *that* diet gives you, *this* diet gives you, too. So find out about them. They'll get you into that enthusiastic frame of mind that's the hallmark of a winner—in dieting and in life.

If you're a vegetarian, this diet is for you. If you're not a vegetarian, you may find dieting easier the vegetarian way. That's because calorie for calorie, there's no difference in the diets, but on the vegetarian version you get those calories from more food.

Doctor, we're talking about a vegetarian diet—a "pure" vegetarian (vegan) diet—no animal foods at all. Is that right?

That's right.

And your diet is based on the right amino acids in the right amounts?

Yes, it is.

But doctor, isn't it awfully hard to get the right amino acids—the essential amino acids—on a pure vegetarian diet?

Ordinarily, yes. But not when you follow my menu plans. They've been carefully worked out to provide *all* the essential amino acids.

How can that be when my nutritionist says there's virtually no plant food that contains all the essential amino acids?

It's true that no plant food contains them all. But, as every nutritionist knows, there's a simple way of getting around that. There are ten essential amino acids. Let's say one plant food contains six of them and another plant food contains the other four. All you have to do is eat both foods at the same meal, and you get all ten essential amino acids.

Are there many plant-food combinations that will give me all the essential amino acids?

There are a good number of these combos. I've prepared a guide for you. It's on the following page. It's not a bad idea to Xerox the guide and carry it with you when you're eating out. For convenience, you can have it miniaturized and carry it in your purse or wallet.

I have a question about beverages. Are you permitting coffee and tea as you did on the nonvegetarian version?

Coffee and tea are really optional on both versions. If you're a coffee or tea drinker, I think it would be an extra burden on you if I asked you to give up the beverage you're used to while you're dieting. Herb teas or diet soft drinks are all right, too. But, please, no sugar.

I have a question about calories. Your First Week on the nonvegetarian diet gives me about 800 calories.

Does the First Week on the vegetarian diet keep calories down that low?

Yes, it does. And because it does, as I explained in my consultations on the nonvegetarian diet, you won't be getting enough of the nutrients you need from your food alone, so the vegetarian diet is supplemented with vitamins and minerals and amino acids, just like the nonvegetarian version.

YOU CAN GET YOUR ESSENTIAL AMINO ACIDS FROM THESE PLANT-FOOD COMBOS

- Legumes plus grains
- Legumes plus seeds
- Legumes plus nuts

Legumes and legume products include dried peas and beans, lentils, peanuts, peanut butter, bean curd (including tofu, which is soybean curd) and soybean milk (use only when fortified).

Examples:
Legumes plus grains

beans and cracked wheat

beans and rice

beans and whole-grain bread

lentils and barley

lentils and rice

tofu and rice

Legumes plus seeds

beans and corn bread

beans and corn tortillas

beans and cornmeal

beans (usually garbanzo) and sesame-seed paste (tahini)

soybeans (roasted), peanuts and sunflower seeds

split peas and corn bread

Legumes plus nuts

garbanzo beans (chick peas) and mixed nuts

soybeans (roasted) and almonds

THE VEGETARIAN DIET (WEEK ONE)

DAY 1

Plain coffee, tea, herb tea or diet soda optional with each meal.

Breakfast

¼ medium	navel orange
¼ cup	enriched farina, with
1 teaspoon	brewer's yeast
½ slice	soy-wheat toast, with
½ teaspoon	margarine
½ cup	fortified soy milk

Lunch

¼ cup	split pea soup, (low sodium) with
2 teaspoons	diced cooked carrot, and
⅛ cup	croutons
2 teaspoons	peanut butter, on
1 slice	whole-grain bread
1	fresh fruit salad, consisting of: • ⅛ cup diced apple • ⅛ cup sliced banana • ¼ cup grapefruit sections • 1 teaspoon sunflower seeds

Dinner

¼ cup	garbanzo beans, boiled, with
	• ¼ cup cooked brown rice
	• ⅛ cup stewed green peppers
	• 1 teaspoon sesame seeds
¼ cup	collard greens, steamed
¼ medium	tomato
1 slice	whole-grain bread, with
½ teaspoon	margarine
¼ medium	pear

Plus 8 glasses of water per day
Don't forget your daily supplements (pages 57–59).

THE VEGETARIAN DIET (WEEK ONE)

DAY 2

Plain coffee, tea, herb tea or diet soda optional with each meal.

Breakfast

1/8 medium	cantaloupe
1/4 cup	bite-size shredded wheat, with
1/4 cup	fortified soy milk, and
1 teaspoon	wheat germ
1 slice	rye toast, with
1/2 teaspoon	margarine

Lunch

1/4 cup	lentil and split pea soup (low sodium)
1/3 cup	garbanzo beans, boiled, sprinkled with
1 teaspoon	sesame seeds
1/2 medium	pita bread, whole wheat
1 small	tomato and cucumber salad, with
1 teaspoon	oil and vinegar dressing
1 medium	plum

Dinner

¼ cup	tofu, heated in nonstick skillet
¼ cup	cooked brown rice
1 small	leafy green salad, with
½ teaspoon	diet mayonnaise
1 small	vegetable salad (your choices)
1 slice	whole-grain bread, with
½ teaspoon	margarine
½ medium	grapefruit
¼ cup	fortified soy milk

Plus 8 glasses of water per day
Don't forget your daily supplements (pages 57–59).

THE VEGETARIAN DIET (WEEK ONE)

DAY 3

Plain coffee, tea, herb tea or diet soda optional with each meal.

Breakfast

½ medium	grapefruit
¼ cup	cream of wheat, regular, with
1 teaspoon	wheat germ
½ slice	whole wheat toast, with
½ teaspoon	margarine
½ cup	fortified soy milk

Lunch

¼ cup	lentil soup (low sodium)
2 teaspoons	peanut butter, spread on
1 medium	pita bread, whole wheat
1 small	leafy green salad, with
1 teaspoon	oil and vinegar dressing
½ cup	strawberries

Dinner

¼ cup	garbanzo beans, cooked with
¼ cup	sliced carrot
¼ cup	tofu (may be heated and added to garbanzo beans and carrots)
1 small	raw vegetable salad (your choices)
1 slice	whole-grain bread, with

½ teaspoon	margarine
½	banana
¼ cup	fortified soy milk

Plus 8 glasses of water per day
Don't forget your daily supplements (pages 57–59).

THE VEGETARIAN DIET (WEEK ONE)

DAY 4

Plain coffee, tea, herb tea or diet soda optional with each meal.

Breakfast

½ medium	grapefruit
¼ cup	bite-size shredded wheat, with
¼ cup	fortified soy milk, and
1 teaspoon	brewer's yeast
½ slice	soy-wheat toast, with
½ teaspoon	margarine

Lunch

¼ cup	lentil soup (low sodium)
⅓ cup	garbanzo beans, boiled, with
1 teaspoon	sesame seeds
2 teaspoons	peanut butter, spread on
1 slice	whole-grain bread
1 small	tomato and cucumber salad, with
1 teaspoon	oil and vinegar dressing
1 small	navel orange

Dinner

¼ cup	tofu, heated in nonstick skillet, with
⅛ cup	crushed almonds
¼ cup	stewed green pepper and tomato

1 small	leafy salad (your choices), with
1 teaspoon	oil and vinegar dressing
1 slice	whole-grain bread, with
½ teaspoon	margarine
½	apple

Plus 8 glasses of water per day
Don't forget your daily supplements (pages 57–59).

THE VEGETARIAN DIET (WEEK ONE)

DAY 5

Plain coffee, tea, herb tea or diet soda optional with each meal.

Breakfast

¼ medium	navel orange
¼ cup	enriched farina, with
1 teaspoon	brewer's yeast
½ slice	soy-wheat toast, with
½ teaspoon	margarine
½ cup	fortified soy milk

Lunch

¼ cup	lentil and split pea soup (low sodium)
⅓ cup	garbanzo beans, boiled, with
1 teaspoon	sesame seeds
½ medium	pita bread, whole wheat
1 small	tomato and cucumber salad, with
1 teaspoon	oil and vinegar dressing
1 medium	peach

Dinner

¼ cup	tofu, heated in nonstick skillet, with
¼ cup	cooked brown rice

1 small	vegetable salad (your choices), with
½ teaspoon	diet mayonnaise
1 slice	whole-grain bread, with
½ teaspoon	margarine
1 small	plum
¼ cup	fortified soy milk

Plus 8 glasses of water per day
Don't forget your daily supplements (pages 57–59).

THE VEGETARIAN DIET (WEEK ONE)

DAY 6

Plain coffee, tea, herb tea or diet soda optional with each meal.

Breakfast

⅛ medium	cantaloupe
¼ cup	bite-size shredded wheat, with
¼ cup	fortified soy milk, and
1 teaspoon	wheat germ
1 slice	rye toast, with
½ teaspoon	margarine

Lunch

¼ cup	split pea soup (low sodium), with
2 teaspoons	diced cooked carrot, and
⅛ cup	croutons
2 teaspoons	peanut butter, spread on
1 slice	whole-grain bread
½ cup	fresh fruit salad (your choices), with
1 teaspoon	sunflower seeds

Dinner

¼ cup	kidney beans, boiled, with
¼ cup	cooked brown rice
½ cup	stewed green peppers and tomatoes

1 slice	whole-grain bread, with
½ teaspoon	margarine
1 small	fruit of your choice

Plus 8 glasses of water per day
Don't forget your daily supplements (pages 57–59).

THE VEGETARIAN DIET (WEEK ONE)

DAY 7

Plain coffee, tea, herb tea or diet soda optional with each meal.

Breakfast

½ medium	grapefruit
¼ cup	cream of wheat, regular, with
1 teaspoon	wheat germ
½ slice	whole wheat toast, with
½ teaspoon	margarine
½ cup	fortified soy milk

Lunch

¼ cup	split pea soup (low sodium), with
½ teaspoon	sesame seeds
⅓ cup	kidney beans, boiled with
3 teaspoons	cooked brown rice
1 small	tomato and cucumber salad, with
1 teaspoon	oil and vinegar dressing

Dinner

¼ cup	garbanzo beans, boiled, with
⅛ cup	crushed nuts of your choice
¼ cup	collard greens, steamed
¼ medium	tomato
1 slice	whole-grain bread, with
½ teaspoon	margarine
¼ medium	peach

Plus 8 glasses of water per day
Don't forget your daily supplements (pages 57–59).

10

The Vegetarian Menu Plan: Week Two Create Your Own Menus for Better Health

Doctor, is your vegetarian diet as healthful as your nonvegetarian diet?

"Most nutritionists agree that vegetarian diets can be adequate if sufficient care is taken in planning them." That quote comes from the Food and Nutrition Board of the National Academy of Sciences. "Sufficient care"—I'll show you how easy it is to take it should you decide to prepare your own menus for Week Two.

If you don't want to prepare your own menus, I've planned Week Two menus with "sufficient care" *for* you. They begin on page 151.

The vegetarian version of The Amino Acid Super Diet is every bit as healthful as the nonvegetarian version.

Doctor, what do you mean by "sufficient care" in planning a vegetarian diet?

First off, to get your full quota of amino acids, you have to combine the right plant foods. You know all about that from our previous consultations. But what about the rest of your nutrient requirements? When you're eating only plant foods, it's not always easy to get all the nutrients you need. But you can do it by making the right combinations of selections from seven groups of plant foods, combinations that supply you with healthful correct amounts of vitamins, minerals, carbohydrates, fats and proteins.

What are the seven groups of plant foods? And why are they healthful?

The seven groups are as follows:

1. High-protein plant foods, which supply amino acids.
2. Dark green and deep orange plant foods, which supply vitamin E and fiber.
3. Plant foods high in vitamin C, which are high in fiber as well.
4. Other fruits and vegetables that are high in fiber and rich in vitamins and minerals.
5. Whole grains and their products, which are rich in B-complex vitamins, iron and fiber.
6. The soybean products tofu and soy milk, which are good amino acid sources.
7. Vegetable oils that are composed almost entirely of healthful polyunsaturated fats and contain no cholesterol. If these oils are cold-pressed, they are also rich in vitamin E.

Do you have lists of the foods in each of the seven groups of plant foods?

Yes. You'll find them beginning on page 145. Select foods according to your taste, and make the most healthful combos by following the instructions on page 144. That's how I made up *my* vegetarian menus.

Doctor, I heard that even if you make selections from the seven groups of plant foods, there are still some nutrient shortages. Is that true?

Yes, but not if you do something about it. I make up for deficiencies in vitamin B_{12}—they could cause anemia—with soy milk fortified with B_{12}. I avoid iron deficiencies—they could also cause anemia—by including iron-rich foods, such as whole grains and broccoli. I provide calcium, which would ordinarily come from cow's milk on a nonvegetarian diet, with such foods as calcium-fortified soy milk and dark green leafy vegetables, which are a fine source of calcium.

On the nonvegetarian diet, you allow 1,000 calories a day during Week Two, 200 more than in Week One. Is it the same for the vegetarian diet?

Yes, and you get the additional calories in a way every dieter loves—from snacks. There's a list of approved ones on page 162.

I like Week One menus. Can I substitute them for Week Two?

Yes. Just slightly increase your portion sizes to pick up the extra 200 calories. Or add the snacks from the list on page 162.

Since a vegetarian diet poses nutrition problems different from those posed by a nonvegetarian diet, must I take different vitamin/mineral and amino acid supplements?

No. The recommended supplements (you'll find them on pages 57–59) are the same on the vegetarian diet as they are on the nonvegetarian diet.

A final question. Weight loss falls off after Week One. Right?

Right.

That puts me in the "danger zone" where I could get discouraged and give up. Do the amino acids keep me as enthusiastic and optimistic as they do on the nonvegetarian diet?

Of course. There's no difference in the amino acids, whether from animals or plants. This vegetarian amino acid diet works equally as well as the nonvegetarian diet.

SEVEN GROUPS OF PLANT FOODS THAT COMBINE FOR A HEALTHFUL DIET

To create your own vegetarian menus for The Amino Acid Super Diet, select the designated number of servings from each of the seven lists of plant foods that follow, and arrange them into breakfast, lunch and dinner menus for each of seven days.

You'll find a complete week's menus created for you, starting on page 151. You can use these instead of making up your own menus; or if you prefer to make up your own menus, you can use them as a model.

LIST 1

High-protein Plant Foods: 4 Servings Daily

1 Serving Is:

¼ cup legumes, cooked

¼ cup tofu

2 teaspoons peanut butter

⅛ cup peanuts, nuts or seeds

Choices:

legumes (dried peas and beans, lentils, peanuts and peanut butter)

nuts

seeds

tofu

LIST 2

Dark Green and Deep Orange Plant Foods: 2 Servings Daily

1 Serving Is:

¼ cup, cooked

½ cup, raw, chopped

1 fruit, small

Choices:

apricot
beet tops
bok choy
broccoli
cantaloupe
carrot
collard greens
dandelion greens
escarole
kale
mango
mustard greens
nectarine
papaya
peach
pumpkin
red pepper
romaine
spinach
sweet potato
turnip greens
winter squash

LIST 3

Plant Foods High in Vitamin C: 3 Servings Daily

1 Serving Is:

¼ cup cooked

½ cup raw, chopped or sliced

1 fruit, small

¼ cup juice

Choices:

cabbage, raw

cauliflower

dark green leafy vegetables

grapefruit and juice

green pepper

guava

mango

orange, or juice

papaya

sprouts (from legumes)

strawberries and other berries

tangerine

tomato, or juice

LIST 4

Other Fruits and Vegetables: 1 Serving Daily

1 Serving Is:

¼ cup cooked

½ cup raw, chopped or sliced

1 fruit, small

Choices:

apple
applesauce
asparagus
banana
beets
celery
cherries
corn
cucumber
grapes
honeydew melon
onion
pear
peas
pineapple
plum
potato, white
prunes
string beans
summer squash
zucchini

LIST 5

Whole Grains and Their Products: 8 Servings Daily

1 Serving Is:

½ slice bread

¼ cup ready-to-eat cereal (nonsugared)

¼ cup cooked grains

¼ cup cooked pasta

1 small cracker (like Wheat Thin)

¼ cup croutons

Choices:

breads

cereals

cornmeal

crackers

granola, without sugar or additives

grits

pasta

rice

LIST 6

Soy Products: 2 Servings Daily

1 Serving Is:

¼ cup milk or tofu

soy milk, fortified

tofu

LIST 7

Vegetable Oils: 1½ Teaspoons Daily

Choices:

corn oil

cottonseed oil

safflower oil

soybean oil

sunflower oil

salad dressings, mayonnaise or margarine made from above oils

THE VEGETARIAN DIET (WEEK TWO)

DAY 1

Plain coffee, tea, herb tea or diet soda optional with each meal.

Breakfast

½ cup	diced apple
¼ cup	nonsugared ready-to-eat cereal (your choice), with
½ cup	fortified soy milk
1 slice	whole wheat bread, with
½ teaspoon	margarine

Lunch

½ cup	bean soup (low sodium), sprinkled with
⅛ cup	sesame seeds
1 slice	whole wheat bread, with
½ teaspoon	margarine
½ cup	cooked broccoli
1 small	tangerine

Dinner

½ cup	tofu, mixed with
2 teaspoons	peanut butter
½ cup	cooked cauliflower
¾ cup	cooked pasta, with
½ teaspoon	margarine

Plus 8 glasses of water per day
Don't forget your daily supplements (pages 57–59).

THE VEGETARIAN DIET (WEEK TWO)

DAY 2

Plain coffee, tea, herb tea or diet soda optional with each meal.

Breakfast

½ cup	sliced strawberries
½ cup	grits, with
½ teaspoon	margarine
½ cup	fortified soy milk, with
1 small	cracker

Lunch

½ cup	broccoli and carrot soup (low sodium), containing ¼ cup each of broccoli and carrot
¼ cup	lentil puree, spread on
1 slice	whole-grain bread
1	fruit salad, consisting of: • ½ cup banana, sliced • ½ cup grapefruit sections • ½ cup orange sections

Dinner

½ cup	garbanzo beans, boiled
1 slice	whole-grain bread, with
1 teaspoon	margarine
2 teaspoons	peanut butter, on
1 small	cracker

Plus 8 glasses of water per day
Don't forget your daily supplements (pages 57–59).

THE VEGETARIAN DIET (WEEK TWO)

DAY 3

Plain coffee, tea, herb tea or diet soda optional with each meal.

Breakfast

1 small	fresh apricot
½ cup	cream of wheat, regular, with
½ teaspoon	margarine
½ cup	fortified soy milk

Lunch

½ cup	split pea soup (low sodium), with
¼ cup	croutons
1 small	salad, consisting of: • ¼ cup raw chopped collard greens • ¼ cup raw chopped cabbage • ¼ cup raw chopped dark green vegetables (your choice) with
1 teaspoon	salad dressing made with sunflower oil
½ cup	grapes

Dinner

¼ cup	garbanzo beans, boiled, with • ½ cup cooked brown rice • ⅛ cup stewed green peppers • 1 teaspoon sesame seeds

1 teaspoon	peanut butter, spread on
1 slice	whole wheat bread
1 medium	mango

Plus 8 glasses of water per day
Don't forget your daily supplements (pages 57–59).

THE VEGETARIAN DIET (WEEK TWO)

DAY 4

Plain coffee, tea, herb tea or diet soda optional with each meal.

Breakfast

¼ cup	orange juice
½ cup	granola (without sugar or other additives), with
½ cup	fortified soy milk
1 slice	whole-grain toast, with
1 teaspoon	margarine

Lunch

½ cup	lentil and split pea soup (low sodium)
1 slice	whole-grain bread
1 cup	mixed vegetables, consisting of • ⅓ cup sliced cooked asparagus • ⅓ cup sliced cooked bok choy • ⅓ cup mashed cooked sweet potato
1 small	pear

Dinner

½ cup	tofu, sliced, sautéed in
2 teaspoons	margarine, and sprinkled with
⅛ cup	sunflower seeds

1 cup	fruit salad, consisting of: • ½ cup chopped papaya • ½ cup chopped mango
1 slice	whole-grain bread

Plus 8 glasses of water per day
Don't forget your daily supplements (pages 57–59).

THE VEGETARIAN DIET (WEEK TWO)

DAY 5

Plain coffee, tea, herb tea or diet soda optional with each meal.

Breakfast

½ cup	applesauce
¼ cup	oatmeal, with
½ cup	fortified soy milk
1 slice	whole wheat bread, with
½ teaspoon	margarine

Lunch

½ cup	vegetable soup (low sodium), combined with
¼ cup	split pea puree
1 slice	whole-grain bread
1 cup	fruit salad, consisting of: • ⅓ cup tangerine sections • ⅓ cup sliced papaya • ⅓ cup sliced banana

Dinner

¼ cup	boiled garbanzo beans, heated with • ¼ cup cooked pasta • ¼ cup stewed red peppers

	• ¼ cup steamed sliced collard greens
1 slice	whole-grain bread, with
1 teaspoon	margarine
1 small	pear

Plus 8 glasses of water per day
Don't forget your daily supplements (pages 57–59).

THE VEGETARIAN DIET (WEEK TWO)

DAY 6

Plain coffee, tea, herb tea or diet soda optional with each meal.

Breakfast

¾ cup	berries in season (your choice)
½ cup	nonsugared ready-to-eat cereal (your choice), with
¼ cup	fortified soy milk
½ slice	whole-grain bread, with
½ teaspoon	margarine

Lunch

½ cup	split pea soup (low sodium)
¾ cup	cooked pasta
1	salad, consisting of: • ½ cup sliced escarole • ¼ cup chopped green pepper • ¼ cup drained bean sprouts with
1 teaspoon	safflower diet mayonnaise

Dinner

½ cup	tofu, steamed, with
¼ cup	cooked brown rice
2 teaspoons	peanut butter on
1 small	cracker
¼ cup	fortified soy milk
1 small	peach or pear

Plus 8 glasses of water per day
Don't forget your daily supplements (pages 57–59).

THE VEGETARIAN DIET (WEEK TWO)

DAY 7

Plain coffee, tea, herb tea or diet soda optional with each meal.

Breakfast

3	stewed prunes
½ cup	oatmeal, with
¼ cup	fortified soy milk
1 slice	whole-grain bread, with
½ teaspoon	margarinc

Lunch

½ cup	cooked garbanzo bean puree, mixed with • ¼ cup raw chopped carrot, and sprinkled with • ⅛ cup sesame seeds
1 slice	whole wheat bread, with
1 small	apple

Dinner

½ cup	tofu, steamed, with
½ cup	cooked brown rice
¼ cup	cooked broccoli
1 small	salad, consisting of chopped dark green leafy vegetables (your choice), with

1 teaspoon	salad dressing made with safflower oil
1 small	piece fresh fruit (your choice)

Plus 8 glasses of water per day
Don't forget your daily supplements (pages 57–59).

HEALTHFUL SNACKS

Select any two snacks for your between-meals enjoyment. Or select any one and double the serving.

apple, or other hand-held fruit	½
graham cracker squares	2
peanut butter	1 teaspoon
popcorn, popped dry, unsalted, unbuttered	1 cup
raisins	⅛ cup
Ry Krisps	2
shelled walnuts	⅛ cup
soy milk, fortified	¼ cup
whole-grain cracker	1

11

The Vegetarian Menu Plan: Week Three The Vegetarian Gourmet, and a Touch of Sugar

Doctor, some vegetarian dishes are delicious. Why can't I have them on a diet?

Sometimes I'm asked, "What is the secret of a successful diet program?" Not the idea behind it, not the concept—but the actual program. In my experience, the answer goes like this:

First, start it off very simply. No frills, no gimmicks. Just a straight low-calorie diet based on a concept—in this case supplying amino acids in the correct amounts. Weight loss is fast, and you win the dieter over to belief in your diet. She wants to stay on it. She's eager to continue. Same thing if the dieter's a he. That's Week One.

Then give the dieter a reward. Let her feel she's not fenced in, that she has to eat such-and-such at such-and-such a time. Dieters resent that. They want freedom of choice—women even more than men—so let them have it. Give them a chance to make up their own menus. If they refuse it, whatever the reason, well, they've had the choice. They're in charge. They like that. They're comfortable with the diet. Give them another reward at the same time: more calories—not much—about 200. But 200 calories added to a low-calorie diet seems like 2,000—and the dieter's satisfaction on the diet goes up. That's Week Two.

Week Three is when you firm up the dieter's will to persevere. You do it with more rewards. Sweets. You *know* what that means to a dieter who's been deprived of them for 2 weeks, 14 days, 336 hours. And then give the dieter a crack at some gourmet food. Remember, overweight people don't get that way because they hate food. They *love* fine food—and here's their opportunity to taste it once again after two solid weeks of diet food. I'm not knocking my diet food. The meals are as interesting and satisfying as they can be. But with a delicious gourmet touch added once a day, that sense of taste deprivation—well, it doesn't exactly vanish—but it does decrease. The dieter knows that I'm on her side, that I'm giving her a program she can live with. She becomes psychologically attached to it. She stays with it. So does he.

If you're like most of my patients. You will, too.

Doctor, I like your program. But won't the sugar in sweets stimulate my appetite and make me put on weight?

Not if it's just a touch of sugar. In a previous consultation I explained how sugar stimulates production of the brain neurotransmitter serotonin. That's the brain substance that makes you feel like you've just finished a great meal. And, of course, it's one of the amino acids on this diet—tryptophan—that produces the serotonin.

When do I eat the sweets?

About 20 minutes before meals. They will make you feel great. And when you sit down to your diet meal, you'll find that your appetite has shrunk. You'll eat your meal with gusto, but you won't want anything more. Of course, all the seven amino acids that help control your weight are working for you, too, but that little touch of sugar can be a mighty big help.

What sweets can I eat?

There's a list on page 102. Select the vegetarian sweets. But because dieting the vegetarian way is—let's face the hard fact—a bit harder than dieting the nonvegetarian way, I'm adding these for you: a small amount—and I do mean a *small* amount—of candy or cake.

The Vegetarian Gourmet—what's that?

That's you when you eat Vegetarian Gourmet Diet dishes—one dish a day.

How do Vegetarian Gourmet Diet dishes fit into the program?

Just make the one indicated substition a day in the menu plan for Week One and it becomes the menu plan for Week Three. For example, turn to Day 2 (page 169), and you'll see, under *Dinner,* that *Tofu Delight* (page 179) has replaced the tofu and brown rice of Day 2, Week One. (The page number is where you'll find the recipe.)

What if I don't know how to cook?

Have somebody in your family who likes to cook do it for you. Or learn to cook—the best way to start is with these fast and easy Vegetarian Gourmet recipes.

You know, if you don't cook or have somebody to cook for you, it's hard to avoid junk foods. And don't think for a minute that there are no vegetarian junk foods. Health food stores are packed with them. Just read the labels. And if you rely on junk foods—crammed with sugar, salt and all the other appetite stimulants—you're going to get fat no matter what. Maybe that's what made you overweight in the first place.

I appreciate what you say, and I'm going to give the cooking you recommend a try. But if it doesn't work out—is there an alternative?

Yes, two as a matter of fact. You can repeat Week One, adding the before-meal sweets and increasing portion sizes slightly. Or you can repeat Week Two, adding the sweets and cutting down on portion sizes slightly. Week Three, with the Vegetarian Gourmet recipes, gives you about 1,000 calories, and you don't want to exceed that.

My very firm suggestion is, you're much better off—for all the reasons I've given you—to add the Vegetarian Gourmet Dishes to your diet this week.

After I've completed Week Three of the Vegetarian diet, but still am not down to my desired weight, what do I do?

As with the nonvegetarian diet, just stay with the Week Three menu plan until you reach your goal weight. For more on lifetime maintenance, see Chapter 12.

THE VEGETARIAN DIET (WEEK THREE)

DAY 1

Plain coffee, tea, herb tea or diet soda optional with each meal.

Breakfast

¼ medium	navel orange
¼ cup	enriched farina, with
1 teaspoon	brewer's yeast
½ slice	soy-wheat toast, with
½ teaspoon	margarine
½ cup	fortified soy milk

Lunch

¼ cup	split pea soup (low sodium), with
2 teaspoons	diced cooked carrot, and

Replace croutons with 3 to 4 Sesame Crunchies (page 178)

2 teaspoons	peanut butter, on
1 slice	whole-grain bread
1	fresh fruit salad, consisting of:

- ⅛ cup diced apple
- ⅛ cup sliced banana
- ¼ cup grapefruit sections
- 1 teaspoon sunflower seeds

Dinner

1/4 cup	garbanzo beans, boiled, with
	• 1/4 cup cooked brown rice
	• 1/8 cup stewed green peppers
	• 1 teaspoon sesame seeds
1/4 cup	collard greens, steamed
1/4 medium	tomato
1 slice	whole-grain bread, with
1/2 teaspoon	margarine
1/4 medium	pear

Plus 8 glasses of water per day
Don't forget your daily supplements (pages 57–59).

THE VEGETARIAN DIET (WEEK THREE)

DAY 2

Plain coffee, tea, herb tea or diet soda optional with each meal.

Breakfast

1/8 medium	cantaloupe
1/4 cup	bite-size shredded wheat, with
1/4 cup	fortified soy milk, and
1 teaspoon	wheat germ
1 slice	rye toast, with
1/2 teaspoon	margarine

Lunch

1/4 cup	lentil and split pea soup (low sodium)
1/3 cup	garbanzo beans, boiled, sprinkled with
1 teaspoon	sesame seeds
1/2 medium	pita bread, whole wheat
1 small	tomato and cucumber salad, with
1 teaspoon	oil and vinegar dressing
1 medium	plum

Dinner

Replace tofu and brown rice with Tofu Delight (page 179)

1 small	leafy green salad, with
½ teaspoon	diet mayonnaise
1 small	vegetable salad (your choices)
1 slice	whole-grain bread, with
½ teaspoon	margarine
½ medium	grapefruit
¼ cup	fortified soy milk

Plus 8 glasses of water per day
Don't forget your daily supplements (pages 57–59).

THE VEGETARIAN DIET (WEEK THREE)

DAY 3

Plain coffee, tea, herb tea or diet soda optional with each meal.

Breakfast

½ medium	grapefruit
¼ cup	cream of wheat, regular, with
1 teaspoon	wheat germ
½ slice	whole wheat toast, with
½ teaspoon	margarine
½ cup	fortified soy milk

Lunch

¼ cup	lentil soup (low sodium)
Replace peanut butter with Sunshine Butter (page 180), on	
1 medium	pita bread, whole wheat
1 small	leafy green salad, with
1 teaspoon	oil and vinegar dressing
½ cup	strawberries

Dinner

¼ cup	garbanzo beans, cooked with
¼ cup	sliced carrot
¼ cup	tofu (may be heated and added to garbanzo beans and carrot)
1 small	raw vegetable salad (your choices)
1 slice	whole-grain bread, with
½ teaspoon	margarine
½	banana
¼ cup	fortified soy milk

Plus 8 glasses of water per day

THE VEGETARIAN DIET (WEEK THREE)

DAY 4

Plain coffee, tea, herb tea or diet soda optional with each meal.

Breakfast

½ medium	grapefruit
¼ cup	bite-size shredded wheat, with
¼ cup	fortified soy milk, and
1 teaspoon	brewer's yeast
½ slice	soy-wheat toast, with
½ teaspoon	margarine

Lunch

¼ cup	lentil soup (low sodium)

Replace garbanzo beans, boiled, and sesame seeds with ⅓ cup Near-Eastern Spread (page 181)

1 medium	pita bread, whole wheat
1 small	tomato and cucumber salad, with
1 teaspoon	oil and vinegar dressing
1 small	navel orange

Dinner

¼ cup	tofu, heated in nonstick skillet, with
⅛ cup	crushed almonds
¼ cup	stewed green pepper and tomato
1 small	leafy salad (your choices), with

1 teaspoon	oil and vinegar dressing
1 slice	whole-grain bread, with
½ teaspoon	margarine
½	apple

Plus 8 glasses of water per day
Don't forget your daily supplements (pages 57–59).

THE VEGETARIAN DIET (WEEK THREE)

DAY 5

Plain coffee, tea, herb tea or diet soda optional with each meal.

Breakfast

¼ medium	navel orange
¼ cup	enriched farina, with
1 teaspoon	brewer's yeast
½ slice	soy-wheat toast, with
½ teaspoon	margarine
½ cup	fortified soy milk

Lunch

¼ cup	lentil and split pea soup (low sodium)

Replace garbanzo beans, boiled, with ⅓ cup Tex-Mex Beans (page 182)

½ medium	pita bread, whole wheat
1 small	tomato and cucumber salad, with
1 teaspoon	oil and vinegar dressing
1 medium	peach

Dinner

¼ cup	tofu, heated in nonstick skillet, with
¼ cup	cooked brown rice
1 small	vegetable salad (your choices), with
½ teaspoon	diet mayonnaise
1 slice	whole-grain bread, with
½ teaspoon	margarine
1 small	plum
¼ cup	fortified soy milk

Plus 8 glasses of water per day
Don't forget your daily supplements (pages 57–59).

THE VEGETARIAN DIET (WEEK THREE)

DAY 6

Plain coffee, tea, herb tea or diet soda optional with each meal.

Breakfast

⅛ medium	cantaloupe
¼ cup	bite-size shredded wheat, with
¼ cup	fortified soy milk, and
1 teaspoon	wheat germ
1 slice	rye toast, with
½ teaspoon	margarine

Lunch

¼ cup	split pea soup (low sodium), with
2 teaspoons	diced cooked carrot, and
⅛ cup	croutons
2 teaspoons	peanut butter, spread on
1 slice	whole-grain bread
½ cup	fresh fruit salad (your choices), with
1 teaspoon	sunflower seeds

Dinner

¼ cup kidney beans, boiled

Replace brown rice with 1 Tortilla (page 183)

½ cup	stewed green peppers and tomatoes
1 slice	whole grain bread, with
½ teaspoon	margarine
1 small	fruit of your choice

Plus 8 glasses of water per day
Don't forget your daily supplements (pages 57–59).

THE VEGETARIAN DIET (WEEK THREE)

DAY 7

Plain coffee, tea, herb tea or diet soda optional with each meal.

Breakfast

½ medium	grapefruit
¼ cup	cream of wheat, regular, with
1 teaspoon	wheat germ
½ slice	whole wheat toast, with
½ teaspoon	margarine
½ cup	fortified soy milk

Lunch

¼ cup	split pea soup (low sodium), with
½ teaspoon	sesame seeds
⅓ cup	kidney beans, boiled with
3 teaspoons	cooked brown rice
1 small	tomato and cucumber salad, with
1 teaspoon	oil and vinegar dressing

Dinner

Replace beans and nuts with Beans With Taco Sauce (page 184)

¼ cup	collard greens, steamed
¼ medium	tomato
1 slice	whole-grain bread, with
½ teaspoon	margarine
¼ medium	peach

Plus 8 glasses of water per day

Don't forget your daily supplements (pages 57–59).

VEGETARIAN GOURMET DISHES

Sesame Crunchies

1½ cups whole wheat flour
¼ cup soy flour
¼ cup unhulled sesame seeds
¼ teaspoon each of salt, ground cinnamon, ground ginger
⅓ cup sunflower oil
½ cup (about) water

1. In large bowl of an electric mixer, combine flours, seeds, salt and spices. Beat in oil on medium speed, a little at a time. Dribble in ⅓ cup water while still beating. Stop mixer.
2. Using hands, scrape dough together, adding sufficient additional water (if necessary) to produce a soft ball. Knead dough until smooth. Preheat oven to 350°F.
3. On a lightly floured board, roll out dough to a thickness of ⅛″. Cut into 2″ circles and place on an ungreased cookie sheet. Bake for 15–20 minutes, until crisp and golden.

Yield: 3–4 dozen crunchies

Tofu Delight

1½ teaspoons safflower oil
½ pound bean curd (tofu), cut into 1" cubes
½ pound combined Chinese cabbage (bok choy), watercress leaves and romaine lettuce, torn into bite-size pieces
2 teaspoons unhulled sesame seeds
2 teaspoons reduced-sodium soy sauce
¼ teaspoon ground ginger
2 dashes cayenne
2 cups hot cooked brown rice

1. In a wok or large iron skillet, heat oil until hot but not smoking. Stir-fry bean curd for 3–4 minutes over medium-high heat. Push to center of cooking vessel.
2. Scatter torn greens around bean curd. Sprinkle with sesame seeds, soy sauce, ginger and cayenne. Cover, and steam for 2 minutes. Uncover, and stir to combine. Serve over cooked brown rice.

Yield: 2–3 servings

Sunshine Butter

¾ cup unsalted peanuts
1 cup shelled sunflower seeds
¼–½ cup peanut or safflower oil
½ teaspoon (or less) salt (optional)

1. Using a food blender or processor, grind nuts and sunflower seeds while slowly adding oil. Add salt, if desired.
2. When butter reaches right consistency, transfer to jar and store in refrigerator, tightly closed.

Yield: About 2 cups

Near-Eastern Spread

⅔ cup garbanzo beans, well rinsed
1 large onion, minced, and sautéed in
2 teaspoons sesame oil
1 teaspoon minced garlic
1½ teaspoons reduced-sodium soy sauce
¼ teaspoon salt (optional)
¼ cup sesame butter (tahini)
½ cup toasted and ground unhulled sesame seeds

1. Cook beans in water to cover until very tender. Drain, reserving liquid. Place beans in bowl.
2. Mash beans, adding a little of the cooking liquid to produce a smooth mixture. Stir in sautéed onion and minced garlic.
3. In another bowl, combine balance of ingredients, and blend. Add to bean mixture, and mix thoroughly. Cover and chill. Use as a dip, spread or filling.

Yield: About 1½ cups

Tex-Mex Beans

1/3 cup margarine
2 cups garbanzo beans, cooked al dente
2 cups mixed unsalted nuts
1/4 teaspoon chili powder
1/4 teaspoon salt (optional)

1. Melt margarine in a medium-sized saucepan. Stir in beans, nuts, chili powder and salt. Preheat oven to 400°F.
2. Grease a 10″ × 15″ × 1″ baking pan. Spread mixture across pan evenly. Bake for 12–15 minutes. Serve warm or chilled.

Yield: Serves 8

Tortillas

2 cups cornmeal flour (not degerminated)
1 cup (about) water

1. Place cornmeal in a large bowl. Add water, a little at a time, and blend with a wooden spoon. When mixture becomes too difficult to work with spoon, use hands to blend well, adding more water (if necessary) to hold dough together.
2. Shape dough into 12 balls. Roll out on a lightly floured board, or press each ball between 2 sheets of wax paper.
3. Bake on hot, lightly greased griddle until lightly browned (about 1½ minutes on each side). Tortillas should be soft and pliable. (Bean filling, page 184, may be spooned onto tortillas. Roll. Top with taco sauce.)

Yield: 12 tortillas

Beans With Taco Sauce

For the sauce:

6 medium tomatoes, chopped
1 cup finely chopped onion
½ teaspoon minced garlic
2 teaspoons dried oregano leaves
1 packet Equal dissolved in 1 teaspoon water (optional)
¼ teaspoon salt (optional)
1 cup red wine vinegar

For the beans:

3 cups cooked kidney beans
½ teaspoon (about) chili powder (season to taste)
⅛ teaspoon cayenne
¼ teaspoon salt (optional)
1 teaspoon oil
½ cup chopped onion
2 teaspoons minced garlic
2 medium fresh tomatoes, skinned and chopped, or ⅔ cup canned tomatoes

1. Combine all sauce ingredients in a bowl. Mix thoroughly, and set aside.
2. In another bowl, combine beans, chili powder, cayenne and salt. Stir to blend, then coarsely mash. Set aside.
3. Heat oil in a large skillet until hot but not smoking. Sauté onion and garlic until wilted. Add tomatoes, and cook 3 minutes while stirring.
4. Add mashed beans to skillet, a little at a time. Bring mixture to simmering point. Cook for 7 minutes. Spoon hot mixture onto tortillas if desired, or serve as separate dish.
5. Top with sauce.

Yield: Enough to fill 12 tortillas

PART IV

Dr. Zisk Solves Your Diet Problems

12

Slim for Life

Doctor, how can I stay slim for life?

I'm going to start this consultation by shocking you a bit. I don't think the table of ideal weights you see in almost every diet book means too much. You see, it's made up by insurance people, and the sample of the population it analyzes is composed of only those who can afford life insurance. I doubt whether that's a fair sample of all the people.

What's more, tables of ideal weights are based on "averages." I don't think averages mean very much. They're unrealistic. The average American family has 2.2 kids. What's two-tenths of a kid? The truth is, nobody is average. We're all different. A famous scientist once said: If you compared the physiological differences between two people of the same sex, you wouldn't think they were two people, you'd think they were two species.

Each of us is born with a certain body frame, and each of us should have enough weight on it to fill it out. That weight is different for every one of us. I call it the "genetically right weight." Everybody has a genetically right weight. *You* have a genetically right weight. Trouble is, before you went on this diet, *you* had *exceeded* yours. And 60,000,000 Americans have exceeded theirs.

What dieting is all about is getting down to your genetically right weight. How do you know when you're there? When you look at your naked self in a full-length mirror and you like what you see. When you and the flesh you carry go together like a hand in a glove. When you feel good about your body. Really good. That's when you know you're down to your genetically right weight.

You know, if more people—particularly young people—took this sensible view of what they should weigh, we would have fewer cases of anorexia, less mental anguish over numbers on a scale, a good deal more happiness, and far more mental energy for going after the good things in life and getting them. Your desired weight, as I've been calling it throughout these consultations, should be your genetically right weight. When you get down to it, stick to it.

But how? The problem is not to get excess weight off, it's to *keep* it off. A prominent weight-control specialist, says that he tells his patients, "Any idiot can *lose* weight. It's *staying* thin that's the test."

Well, I don't think any "idiot" can lose weight. I think it takes a person wise enough to realize that something has to be done for him or her to become more attractive, healthier and more employable—and *does* something. *You* have slimmed down on this diet, and you're no idiot. You were smart to begin with, and smart enough to know that this diet could make you even smarter. How do you feel now that you were smart enough to get down to your desired weight?

"Great!"

Okay, but don't get carried away. The hardest part is still ahead of you. I agree with most M.D.s that *staying* thin is the test. But I have confidence in you. I know you're smart enough to pass it. Up to now, in a group of a hundred people who have lost weight, only five have kept it off after a year. If you join that group of winners, it'll be six. And *you'll* keep it off not for just a year, but always.

Yes, I *can* tell you how to stay slim for life.

Doctor, what do I do right after I get down to my desired weight?

What you've been dreaming about doing all along. Go on a food binge. We're all human, and if I told you not to, you'd do it anyway. But binge just once. After that, stop. Then for a few weeks, stay on the plan for Week Three—with any of the options you choose—but make the portion sizes larger. You don't want to keep losing weight. Watch your scale. Weigh yourself naked before breakfast every day. If you continue to lose weight—and that's normal—eat *more.* If you start to gain weight, cut back. It will take a while, but soon you'll know just how much to eat so your weight won't fluctuate by more than a pound.

Doctor, I loved your diet while I was reducing, but I don't want to stay on it for life. What're the rules for staying slim for life when I'm on my own?

I've made up *Seven Rules for Lifetime Slimness.* They appear on the following page.

SEVEN RULES FOR LIFETIME SLIMNESS

1. *Stay away from junk food,* except for an occasional nibble. Junk food can make you fat.
2. *Stay with natural foods.* As natural as you can get. Home-grown, organically grown, range-fed foods are the best.
3. *Stay away from excess sugar.* It provides calories but no nutrition, so you have to get your nutrition from other foods, and that means eating more calories than you should. Sugar could make you fat.
4. *Stay with low-fat foods.* Remove all visible fat before cooking, and other fat during cooking and at the table.
5. *Stay away from excess salt.* You don't have to be a scientist to know that salt is an appetite stimulant.
6. *Stay with perfect-protein foods.* They supply the amino acids that help you reduce.
7. *Stay with a well-balanced diet,* completely supplemented with amino acids, vitamins and minerals. It provides you with all the nutrients you need, and just the right amount of calories to keep your weight steady.

What's the best way to make up well-balanced menus for lifetime slimness?

Use the instructions I've given you for creating menus from the best foods. For nonvegetarians, the instructions are on page 81; for vegetarians, on page 144.

I'd like to get more gourmet recipes for slim, healthful dishes. Can you recommend a book?

The best book of its kind is Francine Prince's *The Dieter's Gourmet Cookbook: Delicious Low-Fat, Low-Cholesterol Cooking and Baking Recipes Using No Sugar or Salt.* It's an internationally acclaimed best seller. The publisher is Fireside, a division of Simon and Schuster, New York.

Doctor, I'm a person on the run. I don't have time to make up menu plans or cook gourmet foods. Give me a busy person's guide to staying slim for life.

I'll be glad to.

A BUSY PERSON'S GUIDE TO STAYING SLIM FOR LIFE

During the course of a day, *eat* what I tell you to eat and *Avoid* what I tell you to avoid.

Eat perfect proteins—fish, chicken, eggs, meat—never fried—or vegetable "combos" (page 126).

Eat fresh salads and a variety of steamed vegetables.

Eat a variety of fresh fruits.

Eat whole-grain cereals and baked goods.

Eat skim-milk or fortified soy milk products and drink skim milk or no-fat yogurt or fortified soy milk.

Avoid junk foods.

GUIDELINES FOR BETTER EATING-BEHAVIOR PATTERNS

Doctor, I understand that "eating-behavior patterns," if they're bad, can make me put on weight, and if they're good, can keep me slim. What's your advice?

What you say is true.

All of you: Tell me what your eating patterns are, and I'll tell you if they're right or wrong.

Good idea. I eat when I feel like eating, and sometimes I skip meals. Right or wrong?

Wrong. When you eat when you feel like eating, you tend to overeat. When you skip meals, you're hungry as a wolf when you decide to eat, so you overeat. Eat at mealtimes. It doesn't have to be other people's mealtimes. Set up your own schedule. But stick to it. And don't skip meals.

I don't eat breakfast. Right or wrong?

I bet you fill up on all the wrong foods at the coffee break. All that sugar, fat—appetite stimulants. That's the way to being overweight. Stick with the nutritious, slimming breakfasts on The Amino Acid Super Diet. They *control* appetite instead of stimulating it. Eating breakfast—the right kind of breakfast—is a slimming habit.

I like to eat family style: big heaping platters on the table, and everybody serving himself or herself. Right or wrong?

Think about it, and you give me the answer. With all that food out there, isn't there a tempation to eat? And eat? And eat some more? Sure there is. Family-style eating leads to overeating. What you should do is have individual servings of just the right amount of food.

I have individual servings, but I ask for seconds. I know that's wrong, but what can I do about it?

It takes about 20 minutes after you've finished the first servings for your brain to tell your stomach you've had enough. So wait 20 minutes before asking for your seconds, and you won't want them.

I eat in my dining room only. Right or wrong?

Hurray! Somebody's doing something right. If you cultivate the habit of eating in one place only, then you won't nibble in front of the TV, or while you're in your favorite reading chair, or even while you're in bed. Look at all the calories you save.

I keep most of the foods I like out of sight. Way up in the back of closets, or far back in the frig. I figure, out of sight is out of stomach. Right or wrong?

How right you are.

I have better things to do than eating, so I eat fast. Right or wrong?

Wrong. When you eat fast, you're finished before the brain has a chance to send down the signal that you're filled up. You don't know you're filled up, so you eat more, and more and more. The faster you eat, the more you eat. The habit of eating fast, and foods meant to be eaten fast—fast foods—have helped create 60 million overweight Americans.

What can I do to break the fast-eating habit?

First, learn the joys of eating slowly. Gourmets eat slowly, savoring flavors and textures. Of course, the idea of eating fast foods slowly is ridiculous. They're so packed with appetite stimulants that you can't help yourself. So knock fast foods out of your eating pattern. Eating nutritious foods slowly, like a gourmet, is so much more satisfying to your palate. And beneficial to your figure.

I know that at eleven o'clock at night I go to the frig no matter what, as if I'm hypnotized. Don't tell me it's wrong, I know it. Just tell me what to do to break the spell.

You're not alone. Lots of people have "danger times" between meals when they feel compelled to eat. The solution is pretty simple. Just plan to be doing something else at that time that's more attractive.

I have a better solution. I wrap all the foods I love to gorge on during my "danger times" in packages that it takes real effort to open. Funny thing, after struggling with them for about 30 seconds, I give up.

Great idea—for anybody who's tempted to overeat, not only those who are afflicted with "danger times." But some food addicts will struggle until they open the containers. So here's my suggestion: Keep low-calorie foods in the front of the frig—*in easy-to-open containers.* So instead of trying and trying until you open the containers with the high-calorie foods, you'll settle for the low-calorie ones.

Sometimes when things go especially right, I say to the wife, "Hey, let's go out and have a great dinner?" What's wrong with that?

Nothing, if you do it just once in a while. But on a steady basis, those calories mount up. Why not do something else instead to celebrate—go to a movie or a show, go bowling, have some friends over, or just take a nice long walk, hand in hand. The point is this: Never make food a reward.

When I go to a party and the hostess plies me with food, I say "No thanks, I'm watching my weight." Some people think I'm terrible, but I think I'm right. Do you?

I do.

Here's one way I've found to keep no-no foods out of the house. I shop on a full stomach. What do you think of that?

I think it's wonderful. Here's another idea. Should you ever go shopping on an empty stomach—it could happen—prepare a shopping list first. It will guide you past the temptations of the supermarket and stifle impulse buying.

I have the best way in the world to keep my calories down. I skip lunch. Right or wrong?

Oh, so wrong. You eventually arrive at the dinner table ready to eat everything in sight, and you do—at high speed.

Even though I've slimmed down, I think of myself as a fat person. In that way, I'm very conscious of what I eat. Right or wrong?

Some diet doctors might say you're right, but I say you're wrong. Think of yourself as you really are: slim, energetic, younger looking. Then when you're tempted to binge on junk foods, flash that picture in your mind. You won't want to go back to the old fat you. You won't binge.

What troubles me is that a healthful diet is so different from what most of us eat that I feel like a freak. Food editors and TV commercials and food magazines all treat healthful food and healthful diets with contempt. Could they be right?

No. They're not even close to being right. Remember, this diet—or any other healthful diet—is a program for slimming down healthfully and staying slimmed down permanently. Can those food editors and TV commercials promise you that?

STAYING SLIM IN THE REAL WORLD

Doctor, if I were in charge of the world, I'd make everybody eat your way. But I'm not. The world's in charge of me. And sometimes I have to eat their way. How do I stay slim in the real world?

You mean, what do you eat when circumstances are such that it seems you have no choice but to eat the wrong foods?

Yes. Like when the fellers say, "Let's eat at this fast-food joint." I don't want to say no, so what do I do?

You'll just have to make the best of a bad situation. Follow these guidelines. Don't order a superburger with all the trimmings, order a plain burger. Don't order fries, order a baked potato—and eat it plain. Don't order a thick shake, order plain milk; some places have 2 percent milk or skim milk—ask for it. A plain pizza, just bread and cheese, or bread and cheese and vegetables is your dish—not one laden with pepperoni and sausage.

Let's face it. Water is great. But Americans have gotten out of the habit. What do I drink since you say that Coke and all the other regular soft drinks are out?

You have a lot of choices, and often you can find them in vending machines. How about low-fat milk, low-salt tomato juice or V-8? How about seltzer—it comes in many natural flavors now—or club soda, or if you want to be fancy, Perrier or other imported mineral waters? You can make up very interesting drinks for yourself, using fresh fruit juices and any of the carbonated waters.

Okay, but what foods can I get in the supermarket that meet your requirements and are as ready to eat as junk foods?

They may not meet the standards for a totally healthful diet, but if you're in a crunch, these will do: raw vegetables—cut them up in bite-size pieces, of course. Salads with low-calorie, low-salt dressings or plain. Some supermarkets and delis have salad bars where you can pick and choose. What could be easier? How about fresh vegetables steamed rapidly, modern style? Low-sodium soups. Just heat. How about whole-grain breads? They're thick, and chewy, and delicious. How about macaroni? Just takes seconds to make. And you *can* get low-fat crackers.

All right, but how about desserts? Don't tell me you can get really healthful desserts in the supermarket?

Maybe not really healthful. And maybe not as scrumptious as fatty desserts. But the real world is a world of compromise, so here goes. Fruit, plain or with dips. Frozen juice on a stick. Or frozen plain yogurt. Or frozen bananas. Or plain yogurt with just a spoonful of sugarless jam. There are many other desserts like them. And they're not bad tasting at all.

Truth is, with just a little common sense, and with your eyes wide open, you *can* come to terms with the real world while you eat healthfully. All you need is the desire to do it.

13

For Women Only

Doctor, what about those "problem" times in a woman's life?

If you're a man, there are four periods in a woman's life that you can't understand because you're not a woman. If you're a woman, you'll agree. Those periods are menstruation, pregnancy, breast feeding and menopause. Those are trying times for a woman that present all sorts of problems.

I'm a diet doctor. I'll address those problems only in terms of The Amino Acid Super Diet. But some of the boons of this diet seem to solve problems other than those of slimming down and staying slim.

THE PREMENSTRUAL SYNDROME (PMS)

Doctor, I gain 3 pounds before every period. Why me?

You're not alone. You weren't singled out. One out of every three women in this country has the same experience. Let me describe the experience to you and see if I'm not right.

Just before and during menstruation, your ankles, legs and

feet swell? So do your abdomen and your breasts? You feel bloated? Yes? So it isn't just you, is it? As a matter of fact, there are at least 25 million women like you.

Is this a sort of modern epidemic?

Well, medical records on premenstrual syndrome—PMS—which includes weight gain, don't really go back very far, so we don't know if it's modern or not. Some experts suspect it is.

Why do they think PMS is a modern epidemic?

The weight you gain is water—bloat. That means instead of excreting water, you retain it. Now I'm going to speculate.

We eat a great deal of salt in this country. We didn't always. Junk food is responsible. There's salt in everything. Remember the chart (page 69) of all those foods with hidden salt? We can get along nicely on 500 milligrams of salt a day, but we eat up to 11,000 milligrams. Salt holds on to water.

Carbohydrates also hold on to water; and junk-food diets—the nation by and large lives on them—are mostly carbohydrates.

Perhaps the hormonal changes that take place during premenstruation and menstruation trigger some internal mechanism that makes salt and carbohydrates hold on to still more water.

Anyway, we doctors have observed that when patients like you cut down on carbohydrates and salt, they don't become bloated. That's why the Amino Acid Super Diet is a low-salt diet, with just the right amount of carbohydrates. You can continue to lose weight before and during your period, instead of putting it on.

But most American women eat excess salt and carbohydrates. Why does only one out of three suffer from PMS?

We suspect there's a genetic factor. It's the same with high blood pressure. A high salt intake is linked to high blood pressure in some people and not in others. It's genetic. So cutting down on salt may help solve your high blood pressure problem, too, if you have one.

Have you the antidote for other PMS symptoms?

You mean mood swings? *(Yes.)* Insomnia? *(Yes.)* Fatigue? *(Yes.)* Lethargy? *(Yes.)* Depression? *(Yes.)* Nervousness? *(Yes.)* Anxiety? *(Yes.)* Anger? *(Yes.)* Irritability? *(Yes.)*

And cramps, muscular pain and headaches? Can your diet do anything to alleviate those symptoms?

I don't think there is anything we know now that can eradicate PMS, but some things help. You've probably taken pills with aspirin in them to relieve headache and muscular pain; diuretics with caffeine to get rid of water bloat and raise your spirits; and antispasmodics to help curb your cramps. They don't cure, but they help.

The Amino Acid Super Diet can't pull off a miracle, but it can help, too. The diet, which has just the right amounts of carbohydrates and amino acids, helps keep the blood-sugar level from seesawing. An unstable blood-sugar level may be the start of emotional upsets and the headaches that go with them.

Are there any amino acids that specifically act against PMS?

Just one that I know of. Phenylalanine. It's now accepted by doctors as a natural painkiller, because it stimulates the production of the brain's own painkillers, the endorphins. Phenylalanine also produces more of the brain neurotransmitters NE and DA, which raise your spirits—important when fighting the PMS blues.

MENOPAUSE

Doctor, what exactly happens at menopause?

Nature decides your're too old to bear children, so your monthly period stops. It stops because the female hormone estrogen—the biochemical that controls your fertility—ceases to flow as vigorously as before. That results in many changes in your body and in your emotions. That's why menopause is called "change of life."

When does menopause occur?

Between forty-five and fifty-five for most women.

What are the "changes of life" I have to look forward to?

Hot flashes, for one. Sensations of intense heat. They sometimes begin in the chest and spread to the neck and face. Sometimes the heat is all over the body. You'll sweat, but you'll have chills soon after. They're called "flashes" because they last only for seconds or minutes, but they recur again and again. Not pleasant at all.

Can amino acids help hot flashes?

Well, estrogen is not cut off entirely. It's still flowing. And amino acids are the raw materials from which estrogen is made in the body, so you would think that they would help.

Physicians take a more direct route and prescribe drugs that are similar to estrogen. There are about a dozen drugs of this kind, and they are fairly effective. We doctors don't really know, but I suspect that these drugs would be even more effective taken in conjunction with a sound amino acid diet such as mine.

What about osteoporosis at menopause?

Osteoporosis means literally "porous bones." The bones "thin out," and so they're easily fractured. Osteoporosis also shrinks and distorts the skeleton, so you get stooped "little old ladies," and women with "dowager's hump." Treating osteoporosis with estrogen has not been too successful, and I don't see where amino acids enter into the picture at all. But the mineral supplement taken as part of The Amino Acid Super Diet assures an adequate supply of calcium—it's the loss of calcium that's probably the main cause of osteoporosis. And if a calcium supplement is not the whole answer, it may be a help.

Tell me what emotional upsets may be in store for me at menopause, and whether amino acids can help?

The emotional upsets are not unlike the symptoms of PMS. Once again, the hormonal system has changed—the biochemicals in the body are no longer the same as they were—and that affects the brain and nervous system. You may have mood swings, you may have all sorts of emotional problems—it's the time of your life when you can be hard to live with—and there could be depression.

Can amino acids help? Yes, with the emphasis on *help, not* cure. A sound amino acid diet, such as mine, may increase the flow of estrogen, and it's very likely to help keep blood-sugar levels steady. Blood-sugar levels that go up and down like a

Yo-Yo may contribute to mood changes and emotional upset.

Phenylalanine may help with menopausal pain—back pain is common. And phenylalanine and tyrosine, mainly because they help produce the brain neurotransmitters NE and DA, can help beat depression. They could, as a matter of fact, make you feel good. But remember, neither is entirely effective without the support of a complete vitamin/mineral supplement.

Will I gain weight during menopause?

Some women do.

I'd like to stay slim all my life. What advice do you have for a woman entering menopause?

No reason in the world why you can't stay slim. You can reduce on The Amino Acid Super Diet. You can stay thin on its slim-for-life menu plan. And while you're on either, you'll be getting the amino acids, the vitamins and the minerals you need to make menopause more comfortable.

Remember, menopause isn't forever. Some women who come through it, slim and healthy—and free of the biological burden of the monthly cycle—have come to regard postmenopause as "the second spring" of their lives. A sound diet has much to do with that.

THE PILL, PREGNANCY, BREAST FEEDING

Can I reduce while I'm on the Pill?

The argument against most reducing diets while you're on the Pill is this: Low-calorie diets cannot supply you with all the vitamins, minerals, and amino acids you need for good health. The Pill, to make matters worse, diminishes the levels of the following vitamins and minerals in the body: B_1, B_2, B_6, folic acid, bioflavonoids, zinc and chromium.

But on a completely supplemented reducing diet, there are no nutrient deficiencies. The Amino Acid Super Diet is such a diet. And, yes, you can lose weight on it while you're on the Pill.

Can I reduce while I'm pregnant? Or try to keep my weight way down?

No. You would be impairing your baby's health, both mental and physical. Follow your doctor's instructions.

Can I reduce while I'm breast feeding?

If you mean by going on a reducing diet, the answer is no. Your baby is what you feed it. You must eat your full quota of calories to provide the energy and the nutrients your baby needs. Your doctor may also recommend supplements.

On the other hand, breast feeding is a natural way to lose the weight you gained during pregnancy. Breast feeding releases certain substances in your body that get rid of the fat you put on during pregnancy—*fast.* The contours of your uterus, pelvis and waistline rapidly return to normal. Breast feeding is better for your figure than any diet. And it's wonderful for your child.

14

Healthy for Life

Doctor, just how can amino acids improve my health?

I've said before, amino acids are not magic. But they are amazing. They do so many wonderful things. When you come to think of it, that shouldn't surprise anybody. Amino acids are the building blocks of life.

You would expect that on a diet optimally designed to supply the body with amino acids—that is to say, with all the needed amino acids in the right amounts—there would be health benefits. And there certainly are. But don't make the mistake of eating only, or mostly, protein foods to get more and more amino acids so you can get healthier and healthier. It doesn't work that way.

Your body and your brain need fats and carbohydrates, too. And the three essential nutrients must be balanced. They *are* in my diet, with just the right amount of protein—perfect protein that provides you with the optimal amounts of amino acids—and you benefit from all their wonderful health-giving qualities.

I'll answer your questions about just how amino acids improve your health, but keep this in mind: Research on the health effects of amino acids is new, and while the answers I provide are up to the minute, much more research has to be done before the whole truth is nailed down. So before every answer, put these qualifying words, "As of today, it appears that . . ."

Now what would you like to know?

AMINO ACIDS THAT FIGHT DISEASE

Doctor, AIDS is a disease of the immune system. Are there any amino acids that strengthen the immune system?

Arginine seems to stimulate the thymus gland, which produces certain cells—lymphocytes—associated with an efficient immune system. Don't leap to the conclusion that arginine is a cure for AIDS. It's not. But it may help fight off invading organisms, such as viruses, by producing more lymphocytes. AIDS is caused by a virus.

Also, arginine seems to work with another amino acid, ornithine, in a roundabout way to step up production of the growth hormone from the pituitary gland, which in turn acts on the thymus gland to release more T-cells and B-cells. These are powerful virus fighters. So in that roundabout way, arginine *may* be a help against the AIDS virus. But all this is speculation.

Cancer cells are alien invaders of the body. Do arginine and ornithine help fight them?

Only in that these amino acids help increase production of the cells in the immune system that fight cancer cells.

Heart attacks run in my family. Can amino acids help protect me against them?

Two amino acids help the body manufacture a remarkable substance, carnitine, which seems to be in charge of your heart's health. Look what carnitine does:

It reduces the levels of cholesterol and triglycerides in the bloodstream—which most medical experts agree helps ward off atherosclerosis and heart attack. It also helps raise the level of high-density lipoproteins (HDLs)—"the good cholesterol"—which helps prevent cardiovascular disease. And it helps prevent angina—the intense chest pain associated with atherosclerosis and heart attack. It also helps prevent irregular heartbeat.

The two amino acids that go to make up carnitine are lysine and methionine. They are converted into this remarkable substance with the help of the right amounts of vitamins C, B_3, B_6, and the mineral iron.

Is carnitine helpful in other diseases?

There is a group of diseases called "degenerative diseases," all apparently caused at least in part by a diet high in fat and saturated fat, low in fiber, high in calories, and with excessive amounts of sugar and salt. Quite the reverse of The Amino Acid Super Diet. Heart attack and atherosclerosis are degenerative diseases. Carnitine has been tried in the treatment of other degenerative diseases as well—high blood pressure, low blood sugar, also called hypoglycemia, diabetes and diseases of the liver caused by excess fat.

Arthritis is a degenerative disease. I suffer from it. Is there any hope I can be helped by amino acids?

Hope, yes, because some British researchers did report good results when the amino acid cysteine plus a vitamin of the B complex, were used in treating patients suffering from either osteoarthritis or rheumatoid arthritis. Osteoarthritis affects the bones of the joints; rheumatoid arthritis, the soft tissues where the bones of the joints meet. But the treatment has not yet been adopted by American doctors. Cysteine is another form of cystine.

I've read that herpes can be treated successfully with the amino acid lysine. Fact or myth?

The idea behind this belief is that lysine prevents the herpes virus from growing. But there's more to the treatment than that. The amino acid arginine stimulates the growth of the virus. So the treatment removes all foods containing arginine from the diet as well as adding lysine supplements. Herpes simplex I, which causes the common "cold sore," and herpes simplex II, the sexually transmitted viral disease responsible for painful sores in the mouth and on the genitals, are said by some doctors, to be curable by this no-arginine, lots-of-lysine regimen. Other doctors disagree. There's hope here, but not yet a certain cure.

I'm about to be treated at a chronic pain center. The regimen includes dosages of the amino acid phenylalanine. Can it help stop my pain?

Yes, it's standard practice now to prescribe the amino acid phenylalanine, called DLPA, as part of a total pain-control regimen, which includes many other kinds of treatments. Phenylalanine appears to work by stimulating the production of the brain's own painkillers, the endorphins. Tyrosine, an amino acid very much like phenylalanine, is also useful.

Are any other amino acids used to fight chronic pain?

Just one. Not mainly to fight the pain itself, but to help lift the deep depression that almost always accompanies chronic pain. That amino acid is tryptophan. It seems to be effective as part of a total pain-control program. I'd like to point out that none of the amino acids that fight chronic pain are effective unless they are accompanied by adequate quantities of three vitamins of the B complex: B_3, B_6 and inositol.

I suffer from insomnia. I was told that the amino acid tryptophan at bedtime will put me to sleep. Should I try it?

Yes, emphatically. Take it with a glass of milk and it could help solve your sleep problems. It seems to work by stimulating the production of the brain neurotransmitter serotonin, which may be responsible for healthy sleep patterns.

AMINO ACIDS THAT IMPROVE GENERAL HEALTH

We've been wanting to have a child. The doctor says it's not my wife's fault that she hasn't become pregnant. I understand sperm is mostly protein. Can amino acids help?

Nutrition scientists have suspected for a long time that the amino acid arginine is essential for normal sperm production. But fairly recently, a dramatic study proved it. A group of men suffering from low sperm count had been treated with a let's-hope-it-works barrage of medications—vitamins, hormones, even antibiotics. Nothing worked. But after swallowing 4 grams of arginine a day, sperm count shot up, and so did sperm motility in eight out of ten men—and pregnancies followed!

Arginine seems to be a kind of "wonder" amino acid. What else can it do?

It seems to speed healing of wounds, not only from injuries but from surgery. It does this, it appears, by stimulating the production of the growth hormone.

My health and the health of my family are threatened by pollutants. Everybody knows that. Is it possible that amino acids can help fight back?

Cysteine can. It's transformed in the body into a natural protector against pollutants, combining with other amino acids to produce a kind of protein called glutathione, which seems to protect against cigarette smoke—good news for smokers on The Amino Acid Super Diet.

When fats get rancid in the body—and they do when vitamins and minerals are not present in the right amounts—they turn into poisons which sometimes are the basic causes of degenerative diseases. Glutathione protects against those poisons, and also against many others, including the poisons from smog and from alcohol. And if you're a junk-food addict, you'd better add cysteine to your diet—because you need the glutathione made by cysteine to protect you against all those additives.

Can I keep my eyes healthy with amino acids?

Some medical scientists have reported improvements in eye health through sound nutrition. But there's only one possible link with one amino acid. There's a large quantity of taurine in the retina of your eye. Cysteine and methionine combine in your body to manufacture taurine.

Are amino acids linked to healthy skin and hair?

Cysteine helps keep skin and hair looking young and glowing. Cysteine is the major amino acid in hair. You get ample amounts of cysteine on The Amino Acid Super Diet.

Can amino acids extend my life, make me look younger?

That's a tall order. No one really knows. No one's found the elixir of youth. But sound nutrition and a happy, positive state of mind can contribute to looking younger than your real age, and acting younger, too. These days you see a lot of people who follow the rules of the "new nutrition" who look and act so much younger than they are that they shock you when they tell you their real ages. Amino acids play a part in this overall approach to staying young.

Why do we get old? If we knew the answer, what a brave new world this would be! We don't. But we have theories. One is, that we're programmed to sort of start falling apart when we reach a certain age. Almost everything in the body is attacked by poisons it itself produces, called oxidants and free radicals. *They* are the enemies of youth. They turn your skin into wrinkled leather, mark it with age spots, make it look all dried out. They do the same with the inside of your body. They kind of mummify you while you're still alive—and that's the way to old age and death.

We know of only one amino acid that can fight these oxidants and free radicals, hold them off, and slow them down. It's cysteine. It works even better in combination with vitamin E and the mineral selenium. On The Amino Acid Super Diet you get your quota of cysteine.

How will you look five years from now if you stay on The Amino Acid Super Diet? Only four years older, or three, or less? Maybe. Certainly, you'll look youthfully slim.

Index